An Introduction to Mathematical Analysis

An Introduction to Mathematical Analysis

JOHN B. READE

Department of Mathematics
The University, Manchester

CLARENDON PRESS · OXFORD
1986

Oxford University Press, Walton Street, Oxford OX2 6DP

Oxford New York Toronto
Delhi Bombay Calcutta Madras Karachi
Kuala Lumpur Singapore Hong Kong Tokyo
Nairobi Dar es Salaam Cape Town
Melbourne Auckland

and associated companies in
Beirut Berlin Ibadan Nicosia

Published in the United States
by Oxford University Press, New York

British Library Cataloguing in Publication Data

Reade, John B.
An introduction to mathematical analysis.
1. Calculus
I. Title
515 QA303

ISBN 0-19-853258-X
ISBN 0-19-853257-1 Pbk

Library of Congress Cataloging in Publication Data

Reade, John B.
An introduction to mathematical analysis.
Includes index.
1. Mathematical analysis. I. Title.
QA300.R43 1986 515 85-30985

ISBN 0-19-853258-X
ISBN 0-19-853257-1 (pbk.)

Printed in Great Britain by The Universities Press (Belfast) Ltd.

Preface

The aim of this book is to give an introduction to that part of mathematics which has come to be known as *analysis*. It is intended to be read by those who have studied calculus from the point of view of its applications, and are now ready for a deeper analysis of the ideas involved. Prerequisites are a working knowledge of the techniques of calculus, and an ability to manipulate logical arguments. The book is envisaged as suitable for students of mathematics in their first year at university.

Two conflicting objectives are present in every area of mathematics. On the one hand there is the desire to understand the underlying principles. On the other hand there is the need to find answers to practical problems. In its early stages a particular piece of mathematics evolves as a tool for solving certain types of problem. Full understanding comes later. Euclid provided the definitive format for proper understanding of an area of mathematics. The fundamental concepts are those of a *theorem* and a *proof*. Since a proof involves logical deduction of one theorem from others it is necessary to start with theorems which are assumed without proof. Such unproved theorems are called *axioms* which are as few and as simple as possible. Having decided what one's axioms are to be, one is then committed to arguing *rigorously* from these axioms and no others. Euclid applied this rationale to geometry. Mathematical analysis is the application of this rationale to the infinitesimal calculus of Newton and Leibnitz.

Since we aim only to introduce analysis, rather than give a formal treatise on the subject, we have made no attempt at complete coverage. We have concentrated on explaining the basic ideas only, having in mind a student who will read this subject among others, who wants quick access to the essentials, and wishes to avoid being held up on abstruse ramifications. At the same time, we hope that those students who intend to specialize in mathematics, and analysis in particular, will find this book an adequate preparation for any later study they may undertake in this area.

A large number of exercises have been included. Those in the body of the text are straightforward and are meant to confirm an idea that has just been introduced, and no more than that. Those at the end of the chapters are more challenging and are meant to flesh out the material of the particular chapter. A few are results which are to be used later, but were considered unsuitable for inclusion in the main text.

There are a few departures from the standard presentation of analysis at this level. Most notable among these are the emphasis on sequential convergence as the definitive limiting process, and the use of sequences to prove the fundamental theorems about continuous functions. The elementary properties of the exponential and trigonometric functions are obtained without calculus. We give a proof of the Riemann integrability of a continuous function which avoids mention of uniform continuity. Continuity and differentiability of power series are proved as special cases of Weierstrassian theorems about infinite series of functions.

I would like to thank Anthony Watkinson for originally inviting me to write the book, Nicholas Browne for comments from the point of view of English sixth formers, Egbert Dettweiler and Brian Hartley for comments derived from using some of the material in teaching undergraduates, Alan Best, David Brannan, and Phil Rippon for discussions about the sequential approach to continuous functions, Beryl Sweeney for her patience and perseverance in typing the manuscript, and finally my wife Suzanne for her unfailing support and encouragement throughout the whole project.

Manchester
21 December 1984

J. B. R.

Contents

1
The real numbers

Our intention is to apply the rigorous spirit of Euclidean geometry to the subject material of the infinitesimal calculus. We shall postulate a small number of self-evident axioms and then deduce the whole subject logically from these axioms.

Since we shall be dealing with real numbers throughout, we shall present the axioms as a list of properties which we shall assume the real numbers to have. We do not wish to be too pedantic about this. Many properties are so utterly self-evident as to hardly need mentioning. We shall pay particular attention to those properties which may not be quite so familiar to a student embarking on a course in mathematical analysis for the first time.

To this end, we shall take all the *arithmetical* properties of real numbers, such as are concerned with addition, subtraction, multiplication, and division, totally for granted. We shall assume also that the integers and the rational numbers (fractions) are known and that it is unnecessary to define them.

We shall be more careful when it comes to *inequalities*. We shall spell out the axioms for inequalities in detail, and encourage the student at an early stage to obtain as much facility with inequalities as he or she can. This is because we believe that the manipulation of inequalities is at the root of analysis, and that success in analysis is not easy until confidence with inequalities is achieved.

We shall think of the real numbers as the values any continuously varying quantity may take, e.g. mass, length, time, temperature. The values may be positive or negative, and arbitrarily large either way. We shall keep in mind a geometrical picture of the real numbers as laid out along a line called the *real line* which we can imagine as calibrated with the integers as shown below.

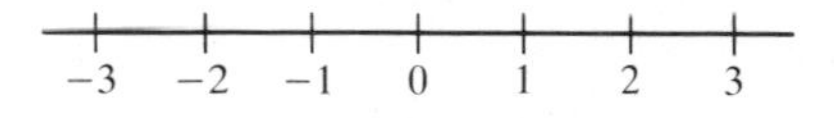

It might be thought that one can adequately describe all the numbers which appear between integer points as rational numbers with suitably large denominators. It turns out that this is not so. In fact, the real number $\sqrt{2}$ cannot be represented as a rational number. This can be easily demonstrated as follows.

Suppose there were integers m, n such that $m^2/n^2 = 2$. We can clearly assume that m, n have no common factor. Multiplying up, we have $m^2 = 2n^2$, from which it follows that m must be even, i.e. $m = 2p$ for some integer p. Substituting for m, we obtain $4p^2 = 2n^2$ which, on cancellation, gives $2p^2 = n^2$. However, this implies that n is even, and therefore m, n have the common factor 2, which is a contradiction. Hence $\sqrt{2}$ must be *irrational.*

1.1 Exercise

Show $\sqrt{3}$, $\sqrt[3]{4}$ are irrational by a similar method. □

As we have already said, we shall assume all the arithmetical properties of real numbers without further ado. We shall also assume the principle of mathematical induction. We give a brief description in case the reader is unfamiliar with it.

Suppose we wish to prove a proposition $P(n)$, with a variable n, is true for all $n = 1, 2, 3, \ldots$ running through the positive integers. Then it is sufficient to prove that $P(1)$ is true, and that, for each $n = 1, 2, 3, \ldots$, $P(n)$ true implies $P(n+1)$ true.

For example, let $P(n)$ be the proposition that the sum of the first n positive integers is $\frac{1}{2}n(n+1)$. $P(1)$ is clearly true, and, if we *assume* $P(n)$ is true for any particular n, then we can deduce $P(n+1)$ is true for this n, since

$$\begin{aligned} 1+2+3+\cdots+n+(n+1) &= \tfrac{1}{2}n(n+1)+(n+1) \\ &= (\tfrac{1}{2}n+1)(n+1) \\ &= \tfrac{1}{2}(n+1)(n+2). \end{aligned}$$

Mathematical induction therefore enables us to conclude that $P(n)$ is true for all positive integers n.

Having disposed of the arithmetical aspects of the real numbers in cavalier fashion, we shall now by contrast concentrate on inequalities in depth.

The real numbers have a natural *order* along the line. Relative position in the order is expressed by saying one real number is *less than* or *greater than* another. We shall use the notation $<$ for less than, and $>$ for greater than, e.g. $2<3$, $5>1$. Expressions involving $<$ or $>$ are called *inequalities.* Inequalities have their own arithmetic which is subject to certain laws in the same way as ordinary arithmetic is. For example, we have the following.

1.2 Law of addition

If $a < b$, then, for any c, we have

$$a + c < b + c. \qquad \square$$

An inescapable consequence of the law of addition is that inequalities for negative numbers may appear at first sight to be the wrong way round. For example, if we start with the inequality $1 < 2$, and subtract 3 from both sides (case $c = -3$), we get $-2 < -1$. This is, however, consistent with the approach of saying $x < y$ if x lies to the left of y on the real line.

1.3 Law of multiplication

If $a < b$, then, for any $c > 0$, we have

$$ac < bc,$$

but, if $c < 0$, we have

$$ac > bc. \qquad \square$$

In words, multiplication of both sides of an inequality by a positive number $c > 0$ preserves the inequality, whilst multiplication by a negative number $c < 0$ *reverses* the inequality. For example, we have $1 < 2$; therefore, multiplying by 3, we get $3 < 6$, or, dividing by 3 (case $c = \frac{1}{3}$), we get $\frac{1}{3} < \frac{2}{3}$, but, if we wish to multiply by -3, we must reverse the inequality and write $-3 > -6$.

There are two other laws which may appear obvious but are none the less important for that.

1.4 Trichotomy law

For any two real numbers a, b one and only one of the three possibilities $a < b$, $a = b$, $a > b$ must occur. $\square$

1.5 Transitive law

If $a < b$ and $b < c$, then $a < c$. $\square$

It follows from 1.4 that, for example, if $a \not> b$ (a is not greater than

b), then either $a < b$ or $a = b$, i.e. a is less than or equal to b, and we write $a \leqslant b$. Similarly, if $a \not< b$, then we must have $a \geqslant b$.

The transitive law 1.5 can be extended to any finite number of terms. For example, if $a < b$, $b < c$, $c < d$, $d < e$, then $a < e$. A standard technique for proving an inequality $a < e$ is to find intervening points b, c, d for which the above chain of inequalities is true.

We shall now illustrate the use of the laws of inequalities just given by *solving* inequalities, which is analogous to solving equations, and *proving* inequalities, which is analogous to proving identities. The rules of course are rather different and do not always lead to the result one might expect. It is essential, however, to abide by the rules at all times. Every step in an argument must be justifiable by reference to one of the four laws given above.

1.6 Worked example

Solve the inequality

$$x + 1 < 2x + 3.$$

Subtracting 1 from both sides gives

$$x < 2x + 2.$$

Subtracting $2x$ from both sides gives

$$-x < 2.$$

Multiplying both sides by -1 gives

$$x > -2.$$

This is the answer. □

1.7 Exercise

Solve

$$\frac{3x - 5}{7} > \frac{2x + 5}{6}.$$ □

1.8 Another worked example

Solve

$$x^2 - 8x + 12 < 0.$$

Method 1 Factorize. We get

$$(x-6)(x-2)<0.$$

Therefore $x-6$, $x-2$ must be of opposite sign. Either $x-6<0$ and $x-2>0$, which gives $x<6$ and $x>2$, i.e. $2<x<6$, or $x-6>0$ and $x-2<0$, which gives $x>6$ and $x<2$, which is impossible. So the answer is $2<x<6$.

Method 2 Complete the square. We get

$$(x-4)^2-4<0.$$

Adding 4 to both sides gives

$$(x-4)^2<4.$$

This inequality can only be satisfied if

$$-2<x-4<2,$$

which, on adding 4 throughout, gives

$$2<x<6. \qquad \square$$

It might be asked how the second argument is justified from the laws of inequality. The argument certainly appeals to common sense, and many might feel this to be sufficient. One has to admit, however, that the reliability of one's common sense depends very much upon the extent of one's experience.

The assumption we have made in this instance is that $x^2<a^2$ is equivalent to $-a<x<a$. This can easily be verified from the graph of $y=x^2$ (Fig. 1.1).

A rigorous justification from the laws of inequality might go like this. If $0<x<a$, then multiplying by x gives $x^2<ax$, and multiplying by a gives $ax<a^2$, and therefore, by the transitive law, $x^2<a^2$.

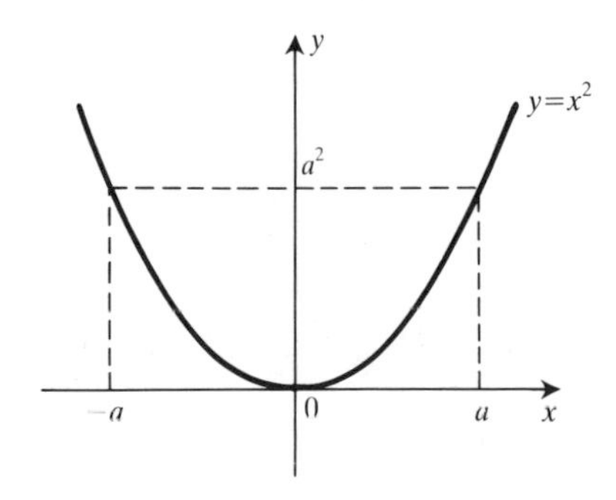

Fig. 1.1

The same argument shows that, if $x > a > 0$, then $x^2 > a^2$. A similar argument shows that $-a < x < 0$ implies $x^2 < a^2$, and that $x < -a < 0$ implies $x^2 > a^2$. This has shown that, if $-a < x < a$, then $x^2 < a^2$, whilst, if $x < -a$ or $x > a$, then $x^2 > a^2$. The equivalence of $x^2 < a^2$ and $-a < x < a$ now follows from the trichotomy law.

1.9 Exercise

Solve

$$x^2 + x - 6 > 0.$$

□

Proving inequalities can be a good deal less straightforward. Many inequalities depend on the fact that $a^2 \geqslant 0$ for all real a. For example, we have the following.

1.10 Worked example

Prove the inequality

$$\left(\frac{a+b}{2}\right)^2 \geqslant ab$$

is true for all real a, b.

Multiplying both sides by 4 gives

$$(a+b)^2 \geqslant 4ab.$$

Subtracting $4ab$ from both sides gives

$$(a+b)^2 - 4ab \geqslant 0,$$

i.e.

$$(a-b)^2 \geqslant 0,$$

which is true, therefore the original inequality is true, since the argument is reversible. □

1.11 Exercise

Prove

$$(ad - bc)^2 \leqslant (a^2 + b^2)(c^2 + d^2)$$

for all real a, b, c, d. □

The transitive law may also come into play as in the following.

1.12 Worked example: Bernoulli's inequality

Prove

$$(1+x)^n \geqslant 1+nx$$

for all real $x>-1$ and all positive integers n.

We argue by induction on n. The inequality clearly holds for $n=1$, in fact is equality for all x. Suppose the inequality is true for a particular n. We shall show this implies it is also true for $n+1$. In fact,

$$\begin{aligned}(1+x)^{n+1} &= (1+x)^n(1+x)\\ &\geqslant (1+nx)(1+x),\end{aligned}$$

since $1+x>0$ on account of the fact that $x>-1$,

$$\begin{aligned}&= 1+(n+1)x+nx^2\\ &\geqslant 1+(n+1)x,\end{aligned}$$

since $nx^2 \geqslant 0$. Therefore, by the transitive law, we have

$$(1+x)^{n+1} \geqslant 1+(n+1)x$$

as required. □

Bernoulli's inequality is of course much easier to prove if we assume $x \geqslant 0$. In fact, we have immediately, from the binomial theorem,

$$\begin{aligned}(1+x)^n &= 1+nx+\tfrac{1}{2}n(n-1)x^2+\cdots+x^n\\ &\geqslant 1+nx\end{aligned}$$

since all the other terms are $\geqslant 0$. Unfortunately this proof fails for $-1<x<0$.

1.13 Exercise

Prove $2^n \geqslant n^2$ for all $n \geqslant 4$. □

Hint Using induction and the transitive law, the problem boils down to showing $2n^2 \geqslant (n+1)^2$ for all $n \geqslant 4$. This can either be proved directly (as in 1.8) or by taking $x=-1/n$, $n=2$ in Bernoulli's inequality. □

An important piece of notation which will be used extensively throughout this book is the so-called *modulus* or *absolute value* of x,

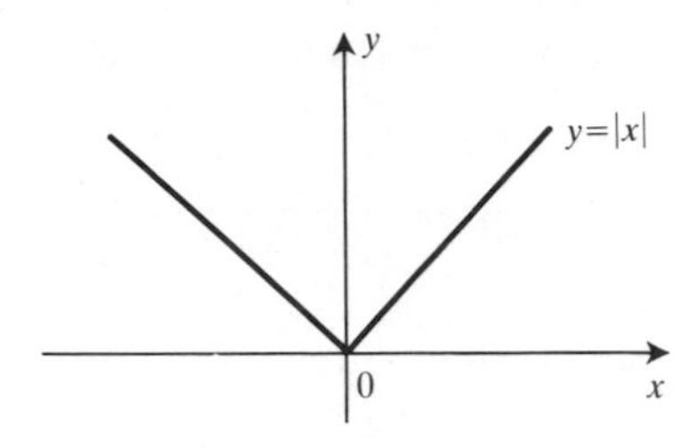

Fig. 1.2

denoted by $|x|$, and defined as follows.

$$|x| = x (x \geqslant 0),$$
$$= -x (x < 0).$$

For example $|-2| = 2$, $|0| = 0$ etc. The graph of $y = |x|$ is as shown in Fig. 1.2.

Observe that $|x| \geqslant 0$ for every x. Also that $|x| = \sqrt{x^2}$ (positive square root) and hence e.g. $|xy| = |x|\,|y|$ for all x, y.

The interaction between modulus and addition is more subtle, and is embodied in an inequality which is important enough to be given the status of a theorem.

1.14 Theorem

For all real x, y we have

$$|x + y| \leqslant |x| + |y|. \qquad \square$$

Proof Squaring both sides gives

$$|x + y|^2 \leqslant (|x| + |y|)^2,$$

which, on expanding and observing that $|x|^2 = x^2$, becomes

$$x^2 + 2xy + y^2 \leqslant x^2 + 2\,|x|\,|y| + y^2,$$

which, on cancelling and using $|xy| = |x|\,|y|$, reduces to

$$xy \leqslant |xy|,$$

which is clearly true. Hence the required inequality follows, since each of the above steps is reversible. $\square$

1.15 Corollary

For all real x, y we have

$$\big||x|-|y|\big| \leqslant |x-y|. \qquad \square$$

Proof Similar to 1.14. □

1.16 Exercises

Prove the following inequalities.

(i) $|ab| \leqslant \frac{1}{2}(a^2+b^2)$.

(ii) $|a+b+c| \leqslant |a|+|b|+|c|$. □

Solving inequalities involving modulus can often be achieved by observing that $|x-y|$ represents the distance between x and y. It follows that, e.g., $|x|<A$ is equivalent to $-A<x<A$, a fact which can itself be used to solve inequalities of certain types.

1.17 Worked example

Solve $|x-4|<7$.

Removing the modulus sign yields

$$-7<x-4<7,$$

which, on adding 4 throughout, becomes

$$-3<x<11,$$

which is the answer. □

1.18 Exercises

Solve the following inequalities.

(i) $|x+1|<1$.

(ii) $|x+2|<|x-2|$. □

All the axioms or laws so far mentioned are satisfied by the rational numbers. And yet the rational numbers do not include $\sqrt{2}$. In order to ensure $\sqrt{2}$ exists as a real number we shall introduce one more axiom called the *upper bound axiom.* Before we can state this axiom, it will be necessary to set up some notation and make some definitions.

1.19 Notation

We shall write $\{x : P(x)\}$, where $P(x)$ is a proposition involving x, to mean the set of all x for which $P(x)$ is true. For example $\{x : x > 0\}$ denotes all positive numbers, $\{1/n : n = 1, 2, 3, \ldots\}$ denotes the set consisting of all reciprocals of positive integers. If $a < b$ are real numbers, we shall write

$$[a, b] = \{x : a \leqslant x \leqslant b\},$$

and call it the *closed interval* from a to b, and

$$(a, b) = \{x : a < x < b\},$$

and call it the *open interval* from a to b. If E is any set and x is any number, we shall write $x \in E$ to mean that x belongs to E. For example, $\frac{1}{2} \in [0, 1]$. □

1.20 Definition

If E is any set of real numbers, and M is another real number, we say M is an *upper bound* of E if $x \leqslant M$ for all $x \in E$. □

1.21 Examples

1 is an upper bound for the closed interval $[0, 1]$. If E is the set of all ages of living American presidents, then 120 is an upper bound for E. □

We define a *lower bound* of E similarly as any m such that $x \geqslant m$ for all $x \in E$. For example, 0 is a lower bound for both sets mentioned in 1.21.

We say E is *bounded above* if E has an upper bound, and *bounded below* if E has a lower bound. We say simply E is *bounded* if E is bounded above and below. For example, both sets of 1.21 are bounded. The set $\{x : x > 0\}$ is bounded below, but not bounded above.

1.22 Definition

We say M is the *maximum* of E, and we write $M = \max E$, if $M \in E$ and $M > x$ for all other $x \in E$, i.e. M is an upper bound of E which belongs to E.

We define the *minimum* of E, denoted by $\min E$, similarly. □

For example, max $[0, 1] = 1$, min $[0, 1] = 0$. However, if $E = \{1/n : n = 1, 2, 3, \ldots\}$, then clearly max $E = 1$, but E has no minimum. This is because no point of E can be a lower bound for E since, for any particular n, we have

$$\frac{1}{n+1} < \frac{1}{n}.$$

1.23 Theorem

Every finite set has a maximum and a minimum. □

Proof This is by induction on the size of the set. If E is the singleton $\{x\}$ consisting of the single point x, then clearly max $E = \min E = x$. Suppose the theorem is true for all sets with n points, and that E has $n + 1$ points. Let

$$E = \{x_1, x_2, \ldots, x_{n+1}\},$$

i.e. E consists of the points $x_1, x_2, \ldots, x_{n+1}$. Then

$$F = \{x_1, x_2, \ldots, x_n\}$$

has n points so has a maximum and a minimum by assumption. Let these be M, m. By the trichotomy axiom, we have $x_{n+1} > M$ or $\leq M$, giving respectively max $E = x_{n+1}$ or M. The argument for min E is similar. □

1.24 Definition

If E is bounded above, then M is the *least upper bound* or *supremum* of E, denoted by sup E, if M is an upper bound of E, and M is less than every other upper bound of E. □

1.25 Examples

Clearly sup $E = \max E$ if max E exists. If E is the open interval $(0, 1)$, then $1 = \sup E$, since clearly 1 is an upper bound, and any other upper bound must be greater than 1. Observe, however, that $(0, 1)$ has no maximum.

1.26 Upper bound axiom

Every non-empty set E of real numbers which is bounded above has a supremum. □

The condition of being non-empty may seem pedantic, but in applications it is essential to know that a bounded-above set E has definitely got something in it before one can deduce the existence of a supremum from the upper bound axiom. Failure to do so can lead to fallacious arguments.

The upper bound axiom is of course very plausible, but it is in fact not at all obvious, indeed it is *false* for the rational number system, as the next theorem shows.

1.27 Theorem

$\sqrt{2}$ exists as a real number. □

Proof Let $E = \{x > 0 : x^2 < 2\}$. Clearly E is non-empty, since e.g. $1 \in E$, and E is bounded above, since e.g. 2 is an upper bound for E. Therefore, by the upper bound axiom, E has a supremum. Let $M = \sup E$. We shall show $M^2 = 2$.

Suppose $M^2 < 2$. Consider $M + 1/n$, where n is a positive integer. Observe that

$$\left(M + \frac{1}{n}\right)^2 = M^2 + \frac{2M}{n} + \frac{1}{n^2}$$

$$\leqslant M^2 + \frac{2M}{n} + \frac{1}{n},$$

for all $n \geqslant 1$,

$$= M^2 + \frac{2M + 1}{n}$$

$$< 2,$$

if

$$n > \frac{2M + 1}{2 - M^2}.$$

Therefore $M + 1/n \in E$ for such n, and of course $M + 1/n > M$, so we have a contradiction of the fact that M is an upper bound of E.

Suppose $M^2 > 2$. Then

$$\left(M - \frac{1}{n}\right)^2 = M^2 - \frac{2M}{n} + \frac{1}{n^2}$$

$$> M^2 - \frac{2M}{n}$$

$$\geqslant 2,$$

if

$$n \geqslant \frac{2M}{M^2 - 2}.$$

Therefore $M - 1/n$ is an upper bound of E for such n, and yet $M - 1/n < M$, which contradicts the fact that M is the *least* upper bound of E.

It follows from the trichotomy axiom that we must have $M^2 = 2$ as required. □

1.28 Corollary

The upper bound axiom is false in the rational number system.

This is because, if it were true, then the argument of 1.27 would show that $\sqrt{2}$ exists in the rational number system, which we know to be not so. □

1.29 Definition

If E is bounded below, we define the *greatest lower bound* or *infimum* of E, denoted by $\inf E$, to be m such that m is a lower bound of E, and m is greater than any other lower bound of E. □

The lower bound axiom then states that any non-empty set E of real numbers which is bounded below has an infimum. We shall take the lower bound axiom as axiomatic, though it can be proved as a consequence of the upper bound axiom (see Exercise 7 at the end of the chapter).

We finish the chapter with a few applications of the upper bound axiom.

1.30 Archimedean axiom

Given any real number x, there exists an integer $n \geqslant x$. □

Another way of saying this is to say that the integers are unbounded above. We tacitly assumed 1.30 in our proof of the existence of $\sqrt{2}$ (see 1.27).

In fact, the Archimedean axiom is a consequence of the upper bound axiom since, if the integers were bounded above, then they would have a supremum M. But then there would have to be an integer n such that $n > M - 1$ (since $M - 1 < M$ and therefore cannot be an upper bound). However, this implies that $n + 1 > M$, which contradicts the fact that M is an upper bound.

1.31 Density of the rationals

We can use 1.30 to show that, between any two real numbers $a < b$, there exists a rational number r, i.e. $a < r < b$.

All we have to do is to choose an integer n such that

$$n > \frac{1}{b - a},$$

and then walk along the real line starting from 0 with steps of length $1/n$.

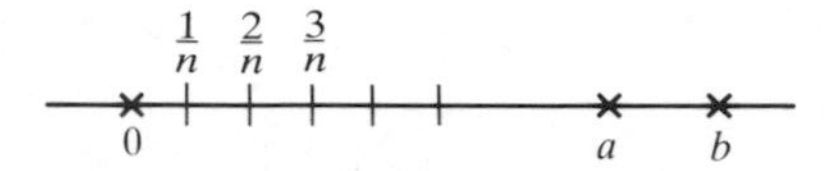

It is clear that we must eventually step into the interval (a, b), i.e. we must have

$$a < \frac{m}{n} < b$$

for some integer m. □

We can show similarly that the irrationals are densely distributed on the real line. To show that there is an irrational number between any two real numbers $a < b$, we take a bus to $\sqrt{2}$ and start walking from there. We must bear in mind, however, that the service may be erratic since Route 2 is irrational!

1.32 Theorem

Any non-empty set of integers which is bounded below has a minimum.

Proof Let the set be E, and let $m = \inf E$. We have to show $m \in E$. Now there must exist $n \in E$ such that $n < m + 1$.

n

m $\qquad$ $m+1$

Therefore $n-1<m$, so there cannot be any integer between n and m. Therefore we must have $n=\min E=m$. □

1.33 Corollary: Mathematical induction

If $P(n)$ is a proposition involving a positive integer n such that $P(1)$ is true, and $P(n)$ true implies $P(n+1)$ true, then $P(n)$ must be true for all n.

Proof If $P(n)$ is false for any n, then the set

$$E=\{n\geqslant 1: P(n) \text{ is false}\}$$

will be non-empty and bounded below. Therefore, by 1.32, E will have a minimum, i.e. there will be a smallest integer n for which $P(n)$ is false. We cannot have $n=1$, since $P(1)$ is true, and we cannot have $n>1$, since then we would have $P(n-1)$ true and $P(n)$ false, which contradicts the induction assumption. We are therefore forced to the conclusion that $P(n)$ is true for all n. □

Observe that any attempt to use 1.23 to prove 1.32 would result in a circular argument since 1.23 was proved by induction!

1.34 Miscellaneous exercises

1. Solve the following inequalities:

(i) $\dfrac{x+1}{x+2}>3$; (ii) $\dfrac{x+1}{x+2}<\dfrac{x+3}{x+4}$;

(iii) $|x+2|>|3x+4|$; (iv) $|2x^2-11x+14|<2$.

2. Prove the following inequalities:
(i) $(a+b)^2\leqslant 2a^2+2b^2$;
(ii) $\sqrt{[(a+b)^2+(c+d)^2]}\leqslant\sqrt{(a^2+c^2)}+\sqrt{(b^2+d^2)}$;
(iii) $||a|-|b||\leqslant|a\pm b|\leqslant|a|+|b|$.
3. Prove $2^n>n^3$ for all integers $n\geqslant 10$.
4. Prove $n!>2^n$ for all integers $n\geqslant 4$.
5. Show that a set E is bounded if and only if there exists M such that $|x|\leqslant M$ for all $x\in E$.
6. Show that if $M=\sup E$, and $\varepsilon>0$ is given, then there exists $x\in E$ such that $x>M-\varepsilon$.

State and prove a corresponding result for $m = \inf E$.

7. Show that the lower bound axiom is a consquence of the upper bound axiom. *Hint:* Given non-empty E bounded below, let F be the set of all lower bounds of E. Show F is non-empty bounded above, therefore has a supremum which must be the infimum of E.

8. Show $\sqrt{3}$, $\sqrt[3]{4}$ exist as real numbers. *Hint:* Argue as in 1.27.

2
Infinite sequences

Infinitesimal calculus is based on the concept of a limit. Differentiation involves taking the limit of dy/dx as dx tends to zero. The definite integral $\int_a^b f(x)\,dx$ is the limit, as dx tends to zero, of finite sums $\sum f(x)\,dx$. A proper understanding of calculus therefore requires analysis of limiting processes.

We shall analyse limiting processes, in the first instance, in the context of convergence of an infinite sequence of real numbers. Having absorbed the basic idea in this relatively simple situation, we shall then be in a position to appreciate more complicated forms of limit.

By an infinite sequence, we mean an unending succession or progression of numbers, e.g. $1, 2, 3, 4, \ldots$, or $2, 4, 6, 8, \ldots$, or $1, \frac{1}{2}, \frac{1}{3}, \frac{1}{4}, \ldots$. We shall use the notation $a_1, a_2, a_3, a_4, \ldots, a_n, \ldots$ for a general sequence. We shall often abbreviate this to $(a_n)_{n\geqslant 1}$, or simply (a_n), or possibly even just a_n, allowing a little innocent confusion between the full sequence $a_1, a_2, a_3, \ldots$ and its nth term a_n.

It may be that we can give a formula for a_n, e.g., $a_n = n$, or $2n$, or $1/n$. It may be that we can't, e.g. $a_n =$ the height in inches of P.C. n of the Nottinghamshire Constabulary.

A sequence will be said to *converge* (or be *convergent*) if its nth term approaches a definite value as n gets larger and larger. For example, the sequence $(1/n)$ converges because $1/n$ gets closer and closer to 0 as n gets larger and larger. Otherwise a sequence will be said to *diverge* (or be *divergent*). For example, the sequence (n) doesn't settle round any value as n gets large.

If the nth term a_n of the sequence (a_n) approaches the value a as n gets large, we shall say (a_n) *converges to a*, and call a the *limit* of (a_n). So, for example, the sequence $(1/n)$ converges to 0.

Notice, however, that $1/n$ is never actually *equal* to 0. All one can say is that $1/n$ *approximates* to 0 to within any required degree of accuracy if n is taken large enough. For example, if accuracy to within 10^{-6} is required, then n must be taken to be greater than 10^6.

This leads us to make the following definition.

2.1 Definition

We say the sequence (a_n) *converges* to the *limit a* if, given any real number $\varepsilon > 0$, there exists a positive integer N such that

$$|a_n - a| < \varepsilon$$

for all $n > N$. □

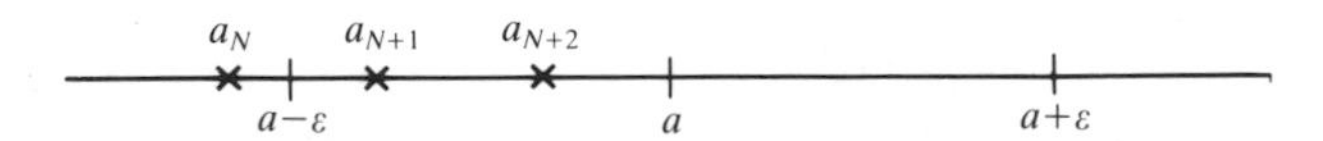

In words, a_n must approximate a to within ε if n is larger than N. Observe that we must allow ourselves to be *given* ε by some notional second person, who is at liberty to choose ε as small as he (or she) pleases, subject only to the condition that $\varepsilon > 0$. Once this second person has committed himself to a particular value for ε, we must then choose N which ensures this degree of accuracy for *every* n beyond N.

It is as if we were playing a kind of game in which one player nominates ε, and the other nominates N. The rules of the game are simply that the ε-player goes first, and the N-player wins if he can find a suitable N, and loses if he can't. In terms of this game, the sequence (a_n) converges to a if the N-player can produce a strategy which gives him a win whatever the ε-player does.

For example, consider the sequence $(1/n)$. We can show $(1/n)$ converges to 0 within the terms of the above definition as follows. If $\varepsilon > 0$ is given, then the strategy of choosing $N \geqslant 1/\varepsilon$ (possible by the Archimedean axiom (1.30)) ensures a win for the N-player since, if $n > N$, then

$$\frac{1}{n} < \frac{1}{N} \leqslant \varepsilon.$$

2.2 Exercises

Give a strategy for choosing N (in terms of ε) such that $a_n < \varepsilon$ for all $n > N$, where a_n is as follows:

(i) $1/n^2$, (ii) $1/2^n$, (iii) $1/\sqrt{n}$. □

Having given a rigorous definition for the limit of a sequence (2.1), we must now construct a theory of convergence with the ultimate object of being able to tell at a glance which sequences converge to which points. We shall concentrate firstly on sequences which converge to zero. This will make the early results easy to state and prove. Later we shall allow sequences to converge to other points. We may even allow them to diverge.

We shall investigate first the interaction between convergence and the operations of arithmetic.

2.3 Definition

We shall call (a_n) a *null sequence* if (a_n) converges to 0. □

Hence, for example, $(1/n)$ is a null sequence.

2.4 Theorem

If (a_n), (b_n) are null sequences, then the sequence $(a_n + b_n)$ formed by adding corresponding terms is also null.

Proof In accordance with the rules of the ε–N game described above, we must firstly allow ourselves to be given $\varepsilon > 0$, and we must then devise a strategy for choosing N which ensures that

$$|a_n + b_n| < \varepsilon$$

for all $n > N$.

Now, we know that (a_n), (b_n) are null, so it follows that N', N'' can be chosen such that $|a_n| < \frac{1}{2}\varepsilon$ for all $n > N'$, and $|b_n| < \frac{1}{2}\varepsilon$ for all $n > N''$. Hence, if we take $N = \max\{N', N''\}$, then we have

$$\begin{aligned} |a_n + b_n| &\leq |a_n| + |b_n| \qquad (1.14) \\ &< \tfrac{1}{2}\varepsilon + \tfrac{1}{2}\varepsilon \\ &= \varepsilon \end{aligned}$$

for all $n > N$. □

2.5 Theorem

If (a_n), (b_n) are null, then so is the sequence $(a_n b_n)$ formed by multiplying corresponding terms.

Proof The proof is similar to that of 2.4. Given $\varepsilon > 0$, we now choose N', N'' such that $|a_n| < \sqrt{\varepsilon}$ for all $n > N'$, and $|b_n| < \sqrt{\varepsilon}$ for all $n > N''$, and take $N = \max\{N', N''\}$. We then have

$$|a_n b_n| = |a_n|\,|b_n| < \sqrt{\varepsilon}\cdot\sqrt{\varepsilon} = \varepsilon$$

for all $n > N$. □

2.6 Definition

The sequence (a_n) is said to be *bounded* if there exists M such that $|a_n| \leqslant M$ for all n. (See 1.34, question 5.) □

For example, $(1/n)$ is bounded, (n) is not.

2.7 Theorem

If (a_n) is null and (b_n) is bounded, then $(a_n b_n)$ is null. □

Proof Let M be such that $|b_n| \leqslant M$ for all n. Then, if $\varepsilon > 0$ is given, choose N such that $|a_n| < \varepsilon/M$ for all $n > N$. It then follows that also

$$|a_n b_n| = |a_n|\,|b_n| < \frac{\varepsilon}{M}\cdot M = \varepsilon$$

for all $n > N$.

2.8 Corollary

If (a_n) is null and C is constant, then (Ca_n) is null. □

Proof Put $b_n = C$ for all n in 2.7. □

2.9 Exercises

Show the following sequences are null.

(i) $\dfrac{1}{n} + \dfrac{2}{n^2} - \dfrac{3}{n^3}$ (ii) $\dfrac{(-1)^n}{n}$ □

Having described the facts of life as they apply to null sequences (2.4 to 2.8), we now proceed to sequences with non-zero limits.

2.10 Theorem

(a_n) converges to a if and only if the sequence $(a_n - a)$ is null. □

Proof This is immediate from the definition of convergence (2.1). □

2.11 Corollaries

If (a_n), (b_n) converge to a, b respectively, then
(i) $(a_n + b_n)$ converges to $a + b$,
(ii) $(a_n - b_n)$ converges to $a - b$,
(iii) $(a_n b_n)$ converges to ab,
(iv) (a_n/b_n) converges to a/b, provided $b_n \neq 0$ for all n, and $b \neq 0$. □

Proofs (i) We have to show the sequence whose nth term is $(a_n + b_n) - (a + b)$ is null. In fact,

$$(a_n + b_n) - (a + b) = (a_n - a) + (b_n - b)$$

so is null by 2.4.
(ii) is similar (use 2.4 and 2.8).
(iii) Observe that

$$a_n b_n - ab = (a_n - a)(b_n - b) + a(b_n - b) + b(a_n - a),$$

so is null by 2.4, 2.5 and 2.8.
(iv) Observe that

$$\frac{a_n}{b_n} = a_n \frac{1}{b_n},$$

so it will be sufficient (by (iii)) to show that $(1/b_n)$ converges to $1/b$. Now

$$\frac{1}{b_n} - \frac{1}{b} = \frac{b - b_n}{b_n b},$$

so (iv) will follow by 2.7 and 2.8, if we can show that the sequence $(1/b_n)$ is bounded.

This can be achieved as follows. We know there must be an N such that

$$|b_n - b| < \tfrac{1}{2}|b|$$

for all $n > N$ (by taking $\varepsilon = \frac{1}{2}|b|$ in the definition 2.1). It follows that

$$\tfrac{1}{2}b < b_n < \tfrac{3}{2}b$$

if $b > 0$, or

$$\tfrac{3}{2}b < b_n < \tfrac{1}{2}b$$

if $b < 0$. The picture for $b > 0$ is as shown in the diagram. Therefore, in all cases, we have $|b_n| > \frac{1}{2}|b|$ for all $n > N$.

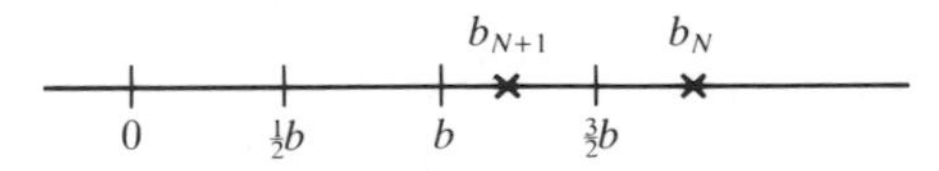

Hence, if we take

$$M = \max\left\{\frac{2}{|b|}, \frac{1}{|b_1|}, \ldots, \frac{1}{|b_N|}\right\},$$

then we have $|1/b_n| \leqslant M$ for all n. □

2.12 Worked example

Find the limit of the sequence whose nth term is

$$\frac{n^2 + 2n + 3}{4n^2 + 5n + 6}.$$

The technique is to write

$$\frac{n^2 + 2n + 3}{4n^2 + 5n + 6} = \frac{1 + 2/n + 3/n^2}{4 + 5/n + 6/n^2},$$

which converges to

$$\frac{1 + 2.0 + 3.0^2}{4 + 5.0 + 6.0^2} = \tfrac{1}{4},$$

by 2.11. □

2.13 Exercises

Find limits of the following sequences.

$$\text{(i)}\ \frac{n+1}{n+2} \qquad \text{(ii)}\ \frac{n+1}{n^2+1} \qquad \text{(iii)}\ \frac{2+n^2}{2-n^2}.$$ □

Our next task will be to consider the interaction between convergence and inequalities.

2.14 Theorem

If $a_n \leqslant b_n$ for all n, and (a_n), (b_n) converge to a, b respectively, then we must have $a \leqslant b$. □

Proof We shall show that the contrary assumption $a > b$ leads to a contradiction. In fact, if $a > b$, then there would exist N', N'' such that

$$|a_n - a| < \tfrac{1}{2}(a-b)$$

for all $n > N'$, and

$$|b_n - b| < \tfrac{1}{2}(a-b)$$

for all $n > N''$. Therefore

$$a_n > \tfrac{1}{2}(a+b) > b_n$$

for all $n > N = \max\{N', N''\}$, which provides the required contradiction. □

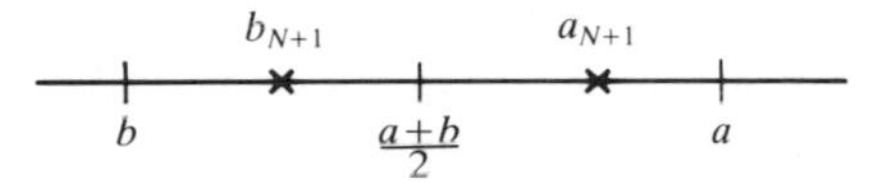

One might expect that if we assume the *strict* inequality $a_n < b_n$ for all n, then we can conclude that strict inequality $a < b$ also holds in the limit. It turns out, however, that we can only conclude that $a \leqslant b$. For example, consider the case $a_n = 0$, $b_n = 1/n$. We have $a_n < b_n$ for all n, but $a = b = 0$.

The moral of the tale is that, if we have an inequality which holds for all values of a positive integer variable n, then it is legitimate to take limits, *provided* we replace $<$ by $\leqslant$ where applicable.

2.15 Theorem: Uniqueness of limits

A sequence (a_n) cannot converge to two different limits $a \neq b$ simultaneously.

Proof If (a_n) converges to a and b, then, taking $a_n = b_n$ in 2.14, we obtain $a \leqslant b$ and $b \leqslant a$. Hence $a = b$. □

2.16 Theorem: Sandwich principle

If (a_n), (b_n), (c_n) are three sequences such that

$$a_n \leqslant b_n \leqslant c_n$$

for all n, and (a_n), (c_n) both converge to the same limit l, then also (b_n) converges to l.

Proof Suppose $\varepsilon > 0$ is given. Then we can choose N', N'' such that

$$|a_n - l| < \varepsilon$$

for all $n > N'$, and

$$|b_n - l| < \varepsilon$$

for all $n > N''$. Therefore, if $N = \max\{N', N''\}$, we have

$$l - \varepsilon < a_n \leqslant b_n \leqslant c_n < l + \varepsilon,$$

and hence

$$|b_n - l| < \varepsilon,$$

for all $n > N$. □

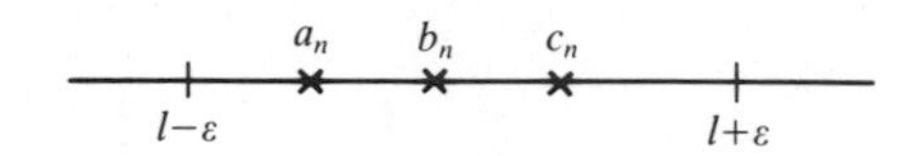

2.17 Application

$(1/2^n)$ is null. In fact, we have

$$0 < \frac{1}{2^n} < \frac{1}{n}$$

for all n, and (0), $(1/n)$ both converge to 0. □

More generally, we can prove the following.

2.18 Theorem

For any fixed x satisfying $|x| < 1$, the sequence $(x^n)_{n \geqslant 1}$ is null. □

Proof
Case 1
$0 < x < 1$

In this case we can write

$$\frac{1}{x} = 1 + h$$

where $h > 0$. Therefore, by Bernoulli's inequality (1.12), we have

$$\frac{1}{x^n} = (1 + h)^n \geqslant 1 + nh,$$

which gives

$$0 < x^n \leqslant \frac{1}{1+nh}.$$

Hence (x^n) is null by the sandwich principle.

Case 2
$-1 < x < 0$

In this case we can write $x = -u$ where $0 < u < 1$. Therefore $x^n = \pm u^n$, and so

$$-u^n \leqslant x^n \leqslant u^n.$$

Hence, again, (x^n) is null by the sandwich principle.

Case 3
$x = 0$

is of course trivial. □

2.19 Exercises

Use the sandwich principle to show that the following sequences are null.

$$\text{(i) } \frac{1}{n!} \qquad \text{(ii) } \frac{2^n + 3^n}{4^n + 5^n}$$ □

We now turn our attention to divergent sequences. There are essentially two ways in which a sequence can diverge. It either 'diverges to infinity' or else it 'oscillates'. The definitive example of divergence to infinity is the sequence (n), and the definitive example of an oscillating sequence is $(-1)^n$.

To say a sequence (a_n) diverges to infinity means that its nth term becomes indefinitely large as n gets large. More precisely, a_n can be made as large as anyone might care to specify, provided we take n large enough. We are again in a two-handed game situation, except that now our opponent nominates a number which can be as *large* as he pleases, and then we have to find N which makes a_n this large for $n > N$. The formal definition is as follows.

2.20 Definition

The sequence (a_n) *diverges to infinity* if, given any $C > 0$, there exists N such that $a_n > C$ for all $n > N$. □

Restricting C to be positive is a matter of convenience, and makes no essential difference since, if N can be chosen corresponding to $C = 1$, say, then this same N will serve equally well for any $C < 1$, including all $C \leqslant 0$.

2.21 Exercises

Find N (in terms of C) such that $a_n > C$ for all $n > N$, where a_n is as follows.

(i) n^2 (ii) $\sqrt{n}$ (iii) 2^n

Hence deduce that the sequences with these nth terms all diverge to infinity. □

One can define a concept of divergence to *minus* infinity in a similar fashion. One simply requires terms to become indefinitely large *negative,* the further one moves along the sequence. The corresponding game involves the first player choosing C as large negative as he pleases, and the second player then choosing N which makes all the terms of the sequence beyond the Nth become larger negative than C. The formal definition is as follows.

2.22 Definition

The sequence (a_n) *diverges to minus infinity* if, given any $C<0$, there exists N such that $a_n < C$ for all $n > N$. □

As for divergence to (plus) infinity, the restriction on C of being negative is for mere convenience, and doesn't affect the game in any essential way.

Examples of sequences which diverge to minus infinity are $(-n)$, $(-n^2)$, $(-\sqrt{n})$ etc. In fact, for any sequence (a_n) which diverges to infinity, the sequence $(-a_n)$ clearly diverges to minus infinity.

Having formally defined convergence and divergence to plus and minus infinity, we shall cover all other cases with the following definition.

2.23 Definition

The sequence (a_n) will be said to *oscillate* if it neither converges, nor diverges to either plus or minus infinity. □

2.24 Example

$(-1)^n$ oscillates.

Proof Suppose $(-1)^n$ were to converge to a. Then there would exist N

such that

$$|(-1)^n - a| < 1$$

for all $n > N$. If we take *even* $n > N$, we obtain

$$|1 - a| < 1,$$

which implies $a > 0$, but, if we take *odd* $n > N$, we obtain

$$|-1 - a| < 1,$$

which implies $a < 0$. So we have a contradiction.

Clearly $(-1)^n$ cannot diverge to plus or to minus infinity. □

Oscillating sequences are in two minds (at least) about what they wish to do. They may try to converge to more than one limit, as in the above example. They may even try to diverge to infinity as well.

There is a theory for sequences which diverge to infinity, analogous to the theory so far developed for convergent sequences. For instance, we have the following.

2.25 Theorem

(i) If (a_n) diverges to infinity and (b_n) is bounded below, then $(a_n + b_n)$ diverges to infinity.

(ii) If (a_n) diverges to infinity and there exists m such that $b_n \geqslant m > 0$ for all n, then $(a_n b_n)$ diverges to infinity.

(iii) *Open sandwich principle* If (a_n) diverges to infinity and $a_n \leqslant b_n$ for all n, then also (b_n) diverges to infinity.

Proofs (i) Suppose $C > 0$ is given. Choose N such that

$$a_n > C - m$$

for all $n > N$, where $m = \inf (b_n)$ (see 1.29). Then

$$a_n + b_n > (C - m) + m = C$$

for all $n > N$.

(ii) If $C > 0$ is given, choose N such that $a_n > C/m$ for all $n > N$. Then

$$a_n b_n > \frac{C}{m} \cdot m = C$$

for all $n > N$.

(iii) Given $C > 0$, choose N such that $a_n > C$ for all $n > N$. Then also $b_n \geqslant a_n > C$ for all $n > N$. □

2.26 Application

$(x^n)_{n\geqslant 1}$ diverges to infinity if $x>1$.

Proof We can write $x=1+h$ where $h>0$. Therefore, by Bernoulli's inequality (1.12), we have

$$x^n=(1+h)^n\geqslant 1+nh,$$

and $(1+nh)_{n\geqslant 1}$ diverges to infinity, by 2.25, (i) and (ii). Hence (x^n) diverges to infinity by 2.25 (iii). □

Observe that for $x<-1$, the sequence (x^n) oscillates. In fact, it tries to diverge to plus and minus infinity simultaneously, a hopelessly ambitious project.

2.27 Worked example

Show that the sequence whose nth term is

$$\frac{n^2+4}{n+4}$$

diverges to infinity.

In fact,

$$\frac{n^2+4}{n+4}\geqslant\frac{n^2}{5n}=\frac{n}{5},$$

so the result follows by 2.25, (ii) and (iii).

2.28 Exercises

Show the following sequences diverge to infinity:

$$\text{(i)}\ \frac{n^3+2}{n^2+2}\qquad\text{(ii)}\ \frac{5^n+4^n}{3^n+2^n}.$$ □

It remains for us to discuss the implications of the upper bound axiom (1.26) in the theory of sequential convergence.

We have already considered bounded sequences (see 2.6), though we have yet to prove one of the most important theorems concerning bounded sequences, an omission we now hasten to repair.

2.29 Theorem

If (a_n) converges, then (a_n) is bounded. □

Proof Suppose (a_n) converges to a. Then there must exist N such that

$$a-1<a_n<a+1$$

for all $n>N$. Therefore, if we take

$$M=\max\{a+1, a_1, a_2, \ldots, a_N\},$$
$$m=\min\{a-1, a_1, a_2, \ldots, a_N\},$$

then we have $m \leqslant a_n \leqslant M$ for all n. □

2.29 has no converse. A bounded sequence (a_n) need not converge. For example, consider the case $a_n=(-1)^n$.

Nevertheless, 2.29 does have certain *restricted* converses, in the sense that one either assumes more or proves less. We shall show that convergence of (a_n) can be proved if we assume a little more about (a_n) than its mere boundedness (see 2.33). On the other hand, we shall also show that, for a general bounded sequence (a_n), even though one cannot prove convergence, one can prove a slightly weaker result, which turns out to be surprisingly useful in applications (see 2.38).

2.30 Definition

A sequence (a_n) is said to *increase* if $a_n \leqslant a_{n+1}$ for all n, or to *decrease* if $a_n \geqslant a_{n+1}$ for all n. (a_n) is said to be *monotonic* if (a_n) either increases (for all n) or decreases (for all n). □

2.31 Examples

$a_n=n$ increases, $a_n=1/n$ decreases but $a_n=(-1)^n$ does neither, i.e., $(-1)^n$ is *not* monotonic. □

2.32 Exercises

Which sequences are monotonic?

(i) $n+(-1)^n$ (ii) $n^2+(-1)^n$ □

2.33 Theorem

If (a_n) is bounded and monotonic, then (a_n) converges. □

This is the first restricted converse of 2.29,

More explicitly, we can isolate the following four possibilities.

(i) If (a_n) increases and is bounded above, then (a_n) converges to its supremum.

(ii) If (a_n) increases but is not bounded above, then (a_n) must diverge to infinity.

(iii) If (a_n) decreases and is bounded below, then (a_n) converges to its infimum.

(iv) If (a_n) decreases and is unbounded below, then (a_n) diverges to minus infinity.

Observe that monotonic sequences never oscillate. They are always single-minded in their intentions.

Proof of 2.33

(i) Suppose that (a_n) increases and is bounded above. Let $M = \sup(a_n)$. We have to show (a_n) converges to M. If $\varepsilon > 0$ is given, then we must have

$$a_N > M - \varepsilon$$

for some N (see 1.34, question 6). Therefore, for all $n > N$, we have

$$M - \varepsilon < a_N \leqslant a_n \leqslant M,$$

and hence

$$|a_n - M| < \varepsilon.$$

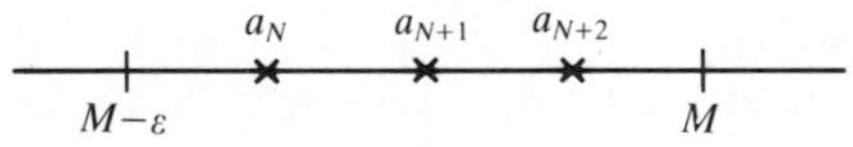

(ii) Suppose now that (a_n) increases and is unbounded above. We now have to show (a_n) diverges to infinity. If $C > 0$ is given, then we must have $a_N > C$ for some N (otherwise C would be an upper bound) and therefore

$$a_n \geqslant a_N > C$$

for all $n > N$.

(iii) and (iv) are proved similarly. □

The beauty of 2.33 is that it enables us to prove a sequence converges without necessarily knowing beforehand what the limit is to be.

2.34 Example

$$a_n = \left(1 + \frac{1}{n}\right)^n.$$

It is not at all obvious what the limit of this sequence might be. We can nevertheless prove it converges by showing it increases and is bounded above.

In fact, expanding a_n by the binomial theorem, we have

$$\begin{aligned} a_n &= 1 + n\frac{1}{n} + \frac{n(n-1)}{2!}\frac{1}{n^2} + \frac{n(n-1)(n-2)}{3!}\frac{1}{n^3} + \cdots \\ &= 1 + 1 + \frac{1}{2!}\left(1 - \frac{1}{n}\right) + \frac{1}{3!}\left(1 - \frac{1}{n}\right)\left(1 - \frac{2}{n}\right) + \cdots, \end{aligned}$$

from which it is clear that a_n increases as n increases. Also that

$$\begin{aligned} a_n &< 1 + 1 + \frac{1}{2!} + \frac{1}{3!} + \cdots + \frac{1}{n!} \\ &< 1 + 1 + \frac{1}{2} + \frac{1}{4} + \cdots + \frac{1}{2^{n-1}} \\ &< 3 \end{aligned}$$

for all n, showing a_n is bounded above. □

In fact, the limit of this sequence is well known and is e, the base of natural logarithms. Indeed, this is one of the ways of *defining* e, though it is not terribly useful in applications, or for numerical calculation of e. Alternative methods of defining e will be considered in Chapter 4.

The second restricted converse of 2.29 is concerned with subsequences, which we now define.

2.35 Definition

If (a_n) is a sequence, then a *subsequence* of (a_n) is any sequence of the form $(a_{n_r})_{r\geq 1}$, where $(n_r)_{r\geq 1}$ is a *strictly* increasing sequence of positive integers, i.e. $n_r < n_{r+1}$ for all r. □

2.36 Examples

(i) $n_r = 2r$. This gives the subsequence $a_2, a_4, a_6, a_8, \ldots$, which can be alternatively denoted by $(a_{2n})_{n\geq 1}$, and is called the *even* subsequence of (a_n).

(ii) $n_r = 2r - 1$. This gives the *odd* subsequence a_1, a_3, a_5, $a_7, \ldots$, alternatively denoted by $(a_{2n-1})_{n\geq 1}$.

Other subsequences we shall have occasion to consider in Chapter 3 are the following.

(iii) $(a_{2^n})_{n\geq 1} = a_2, a_4, a_8, a_{16}, \ldots$.

(iv) $(a_{n^2})_{n\geq 1} = a_1, a_4, a_9, a_{16}, \ldots$. □

2.37 Theorem

If $(a_n)_{n\geq 1}$ converges to a limit a, or diverges to plus or to minus infinity, then any subsequence $(a_{n_r})_{r\geq 1}$ does the same. □

Proof Suppose $(a_n)_{n\geq 1}$ converges to a. Then, for any given $\varepsilon > 0$, we can choose N such that

$$|a_n - a| < \varepsilon$$

for all $n > N$. Now choose R such that $n_r > N$ for all $r > R$. This is clearly possible since n_r increases strictly with r. It then follows that

$$|a_{n_r} - a| < \varepsilon$$

for all $r > R$, and hence $(a_{n_r})_{r\geq 1}$ converges to a.

If $(a_n)_{n\geq 1}$ diverges to infinity, then, for any given $C > 0$, we can choose N such that $a_n > C$ for all $n > N$, and then R such that $n_r > N$ for all $r > R$. Hence $a_{n_r} > C$ for all $r > R$, which proves $(a_{n_r})_{r\geq 1}$ diverges to infinity.

Divergence to minus infinity is treated similarly. □

Observe that single-minded sequences carry all their subsequences with them. Oscillatory sequences have no such power. Consider, for example, the sequence $(-1)^n$. Its even subsequence converges to 1, whilst its odd subsequence converges to -1.

This, incidentally, gives an alternative proof of the divergence of $(-1)^n$ (in the sense of non-convergence), since, if $(-1)^n$ were to converge to a limit l, then, by 2.37, all its subsequences would converge to l, and therefore we would have $l = 1 = -1$, which contradicts the uniqueness of limits (2.15).

We can now present our second restricted converse to 2.29.

2.38 Bolzano–Weierstrass theorem

Any bounded real sequence $(a_n)_{n\geq 1}$ must have a convergent subsequence $(a_{n_r})_{r\geq 1}$. □

Observe that the bounded sequence $(-1)^n$, even though not convergent, does possess convergent subsequences, e.g. the even and odd subsequences.

Proof of 2.38 Since (a_n) is bounded, there exist $A < B$ such that $A \leqslant a_n \leqslant B$ for all n. If we bisect the closed interval $[A, B]$, at least one half must contain a_n for infinitely many n. Let $[A_1, B_1]$ be such a half. If we now bisect $[A_1, B_1]$, then, again, at least one half must contain a_n for infinitely many n. Call such a half $[A_2, B_2]$. We can repeat this process indefinitely to obtain a nested sequence $[A_r, B_r]_{r\geqslant 1}$ of closed subintervals of $[A, B]$ with the following properties.

(i) $A \leqslant A_1 \leqslant A_2 \leqslant \cdots \leqslant A_r < B_r \leqslant \cdots \leqslant B_2 \leqslant B_1 \leqslant B$.

(ii) $B_r - A_r = \dfrac{B-A}{2^r}$.

(iii) For every r, $[A_r, B_r]$ contains a_n for infinitely many n.

For example, the sequence might begin as shown below.

A_1 B_1

A A_2 B_2 B

Observe that $(A_r)_{r\geqslant 1}$ is increasing, bounded above, therefore converges to a limit l, and that $(B_r)_{r\geqslant 1}$ is decreasing, bounded below, therefore converges to a limit l'. (See 2.33, (i) and (iii).) Observe further, that taking limits in (ii) gives $l' - l = 0$, i.e. $l = l'$.

So, to construct a convergent subsequence $(a_{n_r})_{r\geqslant 1}$ of (a_n), we simply choose $a_{n_r} \in [A_r, B_r]$, for each r, in such a way that n_r strictly increases with r. Condition (iii) clearly ensures that this is possible. We then have

$$A_r \leqslant a_{n_r} \leqslant B_r$$

for all r, and therefore, by the sandwich principle (2.16), we must have $(a_{n_r})_{r\geqslant 1}$ convergent to l. □

The Bolzano–Weierstrass theorem will be seen in later chapters to have important consequences in the development of analysis, in particular the minimax theorem for continuous functions (see 4.58), which is essential for Rolle's theorem (see 5.15) and the mean value theorem (see 5.16) in the differential calculus.

2.39 Miscellaneous exercises

1. Discuss the convergence or divergence of the following sequences.

(i) $\left(1+\frac{1}{n}\right)^5$ (ii) $(1+\frac{1}{5})^n$ (iii) $\left(1+\frac{1}{n}\right)^{n^2}$

(iv) $\left(1+\frac{1}{n^2}\right)^n$ (v) $\left(\frac{1}{2}+\frac{1}{n}\right)^n$ (vi) $\frac{2^n+3^n}{3^n+4^n}$

(vii) $n+(-1)^n$ (viii) $n+(-1)^n n$ (ix) $n+(-1)^n n^2$

2. Prove that if (a_n) converges to a, then $(|a_n|)$ converges to $|a|$. (Use 1.15.)

3. Prove the inequality

$$|\sqrt{a}-\sqrt{b}| \leqslant \sqrt{|a-b|}$$

for all $a \geqslant 0$, $b \geqslant 0$.

Deduce that, if (a_n) converges to a, where $a_n \geqslant 0$ for all n, then $(\sqrt{a_n})$ converges to $\sqrt{a}$.

Find the limit of the sequence whose nth term is $\sqrt{(n+1)}-\sqrt{n}$.

4. Prove that, if (a_n) converges to a, then any sequence obtained from (a_n) by altering a finite number of terms also converges to a.

Deduce the following generalized version of the sandwich principle. If $a_n \leqslant b_n \leqslant c_n$ for all $n \geqslant$ some N, and if (a_n), (c_n) both converge to l, then also (b_n) converges to l.

Show the sequence $(n/2^n)$ is null. (Use 1.13.)

5. Show that, if $a_n \neq 0$ for all n, and if (a_n) diverges to infinity, then $(1/a_n)$ is null.

Is there a converse?

6. Suppose (a_n) is an increasing sequence.

Show that, if (a_n) has a subsequence $(a_{n_r})_{r\geqslant 1}$ which converges to a limit a, then (a_n) must itself converge to a. (Use 2.33 and 2.37.)

Show, on the other hand, that, if (a_n) has a subsequence which diverges to infinity, then (a_n) must diverge to infinity.

7. Given that $(1+1/n)^n$ converges to e, show that

(i) $\left(1-\frac{1}{n}\right)^n$ converges to $1/\mathrm{e}$,

(ii) $\left(1+\frac{2}{n}\right)^n$ converges to e^2.

Hints: (i) Observe that

$$1-\frac{1}{n}=\frac{1}{1+\dfrac{1}{n-1}}.$$

(ii) Show $(1+2/n)^n$ is increasing, and that its even subsequence converges to e^2. Now use question 6.

8. Show that, if the set E is bounded above, then one can find a sequence of points $a_n \in E$ which converges to $M = \sup E$. (Use 1.34, question 6.)

Show also that there is a sequence of points $b_n \notin E$ which converges to M.

9. Show that, if E is unbounded above, then there is a sequence of points $a_n \in E$ which diverges to infinity.

10. Show that, if the sequence (a_n) is bounded and divergent, then it has two subsequences which converge to distinct limits. *Hint*: Use the Bolzano–Weierstrass theorem (2.38) twice.

11. Let $A > 0$ be fixed, and let the sequence (a_n) be defined inductively as follows.

$$a_1 = 1,$$

$$a_{n+1} = \tfrac{1}{2}\left(a_n + \frac{A}{a_n}\right)$$

if $n \geqslant 1$. Show that

(i) $a_n^2 \geqslant A$ for all $n \geqslant 2$,

(ii) a_n decreases for all $n \geqslant 2$.

Deduce that (a_n) converges to a limit a.

Find the value of a by taking limits in the defining equation for (a_n).

12. Show that, for any $0 < a < b$,

$$a < \frac{2ab}{a+b} < \frac{a+b}{2} < b.$$

Given $0 < a_1 < b_1$, let sequences (a_n), (b_n) be defined by saying

$$a_{n+1} = \frac{2a_n b_n}{a_n + b_n}, \qquad b_{n+1} = \frac{a_n + b_n}{2}$$

for all $n \geqslant 1$. Prove (a_n), (b_n) both converge to a common limit l.

Find the value of l when $a_1 = 1$, $b_1 = 2$.

3
Infinite series

In common parlance the words 'sequence' and 'series' mean much the same thing. In mathematics, however, they have different meanings. Whereas a *sequence* is a succession of numbers,

$$a_1, a_2, a_3, \ldots, a_n, \ldots,$$

a *series* is a succession of numbers which are supposed to be added together,

$$a_1 + a_2 + a_3 + \cdots + a_n + \cdots.$$

An *infinite series* simply means an *infinite sum.*

To get a precise definition of an infinite sum, we have to regard it as a limit of finite sums. If we write

$$s_n = a_1 + a_2 + \cdots + a_n$$

for each n, and if the *sequence* $(s_n)_{n\geqslant 1}$ converges to a limit s, then we say the infinite sum

$$a_1 + a_2 + \cdots + a_n + \cdots$$

has the value s.

For example, suppose $a_n = 1/2^n$. Then

$$s_n = \tfrac{1}{2} + \tfrac{1}{4} + \cdots + \frac{1}{2^n} = 1 - \frac{1}{2^n}$$

which converges to 1. So we write

$$\tfrac{1}{2} + \tfrac{1}{4} + \cdots + \frac{1}{2^n} + \cdots = 1.$$

Infinite series occur at several strategic points in the development of analysis. For instance, Taylor's theorem in the differential calculus says that, under suitable conditions, any function $f(x)$ can be expanded as

$$f(x) = f(a) + (x-a)f'(a) + \tfrac{1}{2}(x-a)^2 f''(a) + \cdots,$$

where $f'(a)$, $f''(a)$ are successive derivatives of $f(x)$ evaluated at

$x = a$. Important cases are

$$e^x = 1 + x + \frac{x^2}{2!} + \cdots,$$

$$\sin x = x - \frac{x^3}{3!} + \frac{x^5}{5!} - \cdots,$$

$$\cos x = 1 - \frac{x^2}{2!} + \frac{x^4}{4!} - \cdots,$$

all of which are actually called *Maclaurin* expansions by a quirk of language. Not only are these expansions useful for approximating the functions in question, they also provide us with a completely rigorous method of *defining* these functions which is not dependent on geometrical intuition. (See Chapter 4.)

It is therefore relevant to make a rigorous analysis of infinite series at an early stage, and the weapons for achieving this are ready to hand in the form of the theory of convergence of infinite sequences as developed in Chapter 2.

Before making the necessary definitions and commencing the theory, it will be useful to set up the summation notation. We shall use the following abbreviations.

$$\sum_{n=1}^{N} a_n = a_1 + a_2 + \cdots + a_N.$$

$$\sum_{n=1}^{\infty} a_n = a_1 + a_2 + \cdots + a_n + \cdots.$$

We call $\sum_{n=1}^{N} a_n$ the *Nth partial* sum of the *infinite series* $\sum_{n=1}^{\infty} a_n$. We call a_n the *nth term*. We shall often trim down the notation to $\sum_1^N a_n$, $\sum_1^\infty a_n$, and write $s_N = \sum_1^N a_n$, $s = \sum_1^\infty a_n$.

3.1 Definition

We say the infinite series $\sum_1^\infty a_n$ *converges* (or is *convergent*) if the infinite sequence $(s_N)_{N\geq 1}$ of Nth partial sums converges. We call the limit s of $(s_N)_{N\geq 1}$, if it exists, the *sum* of the series $\sum_1^\infty a_n$, and also denote it by $\sum_1^\infty a_n$. □

For example, $\sum_1^\infty 1/2^n$ converges and its sum is 1, and we also write $\sum_1^\infty 1/2^n = 1$.

In other cases we describe the behaviour of $\sum_1^\infty a_n$ by saying what the behaviour of $(s_N)_{N\geq 1}$ is, e.g. $\sum_1^\infty a_n$ diverges to infinity, oscillates, etc.

3.2 Example: Geometric series

$\sum_1^\infty x^n$ where x is fixed.

Here we have

$$s_N = x + x^2 + \cdots + x^N.$$

Therefore

$$xs_N = x^2 + x^3 + \cdots + x^{N+1}.$$

Subtracting, we obtain

$$(1-x)s_N = x - x^{N+1},$$

which gives

$$s_N = \frac{x - x^{N+1}}{1-x},$$

provided $x \neq 1$. Clearly $s_N = N$ if $x = 1$.

It follows that (s_N) converges if $|x| < 1$, and diverges if $|x| \geqslant 1$ (see 2.18 and 2.26). In fact,

$$\sum_1^\infty x^n = \frac{x}{1-x}$$

if $|x| < 1$, whereas $\sum_1^\infty x^n$ diverges to infinity if $x \geqslant 1$ and oscillates if $x \leqslant -1$. □

3.3 Exercises

(i) Show that the Nth partial sum of the series $\sum_1^\infty 1/n(n+1)$ is

$$\sum_1^N \frac{1}{n(n+1)} = \frac{N}{N+1}.$$

Hint: Use induction, or resolve $1/n(n+1)$ into partial fractions.

(ii) Deduce that the series $\sum_1^\infty 1/n(n+1)$ converges, and find its sum. □

3.4 Example: The harmonic series

$\sum_1^\infty 1/n$. Here we have

$$s_N = 1 + \tfrac{1}{2} + \tfrac{1}{3} + \cdots + \frac{1}{N}$$

for which there is no simple formula which enables the convergence or otherwise of (s_N) to be quickly decided. However, we can say

$$s_{2^N} = 1 + \tfrac{1}{2} + (\tfrac{1}{3} + \tfrac{1}{4}) + (\tfrac{1}{5} + \tfrac{1}{6} + \tfrac{1}{7} + \tfrac{1}{8}) + \cdots + \left(\frac{1}{2^{N-1}+1} + \cdots + \frac{1}{2^N}\right)$$

$$> 1 + \tfrac{1}{2} + (\tfrac{1}{4} + \tfrac{1}{4}) + (\tfrac{1}{8} + \tfrac{1}{8} + \tfrac{1}{8} + \tfrac{1}{8}) + \cdots + \left(\frac{1}{2^N} + \cdots + \frac{1}{2^N}\right)$$

$$= 1 + \tfrac{1}{2} + \tfrac{1}{2} + \tfrac{1}{2} + \cdots + \tfrac{1}{2}$$

$$= 1 + \tfrac{1}{2}N.$$

Therefore, by the open sandwich principle (2.25), it follows that $(s_{2^N})_{N\geq 1}$ diverges to infinity. Now, $(s_N)_{N\geq 1}$ is monotonic increasing, so must either converge or diverge to infinity (see 2.33). If $(s_N)_{N\geq 1}$ were to converge, then $(s_{2^N})_{N\geq 1}$, being a subsequence, would also converge (see 2.37). But we have just shown $(s_{2^N})_{N\geq 1}$ diverges to infinity. Hence $(s_N)_{N\geq 1}$ must diverge to infinity, i.e., $\sum_1^\infty 1/n$ diverges to infinity. □

3.5 Exercise

Show by a similar method that the series $\sum_1^\infty 1/\sqrt{n}$ diverges. □

3.6 Example: Euler's series

$\sum_1^\infty 1/n^2$. This time we have

$$s_N = 1 + \frac{1}{2^2} + \frac{1}{3^2} + \cdots + \frac{1}{N^2}$$

for which there is again no simple formula. However, the method of 3.4 can again be applied, but now with the inequalities reversed, to give

$$s_{2^N} = 1 + \frac{1}{2^2} + \left(\frac{1}{3^2} + \frac{1}{4^2}\right) + \left(\frac{1}{5^2} + \frac{1}{6^2} + \frac{1}{7^2} + \frac{1}{8^2}\right)$$

$$+ \cdots + \left(\frac{1}{(2^{N-1}+1)^2} + \cdots + \frac{1}{(2^N)^2}\right)$$

$$< 1 + \frac{1}{2^2} + \left(\frac{1}{2^2} + \frac{1}{2^2}\right) + \left(\frac{1}{4^2} + \frac{1}{4^2} + \frac{1}{4^2} + \frac{1}{4^2}\right)$$

$$+ \cdots + \left(\frac{1}{(2^{N-1})^2} + \cdots + \frac{1}{(2^{N-1})^2}\right)$$

$$= 1 + \tfrac{1}{4} + \tfrac{1}{2} + \tfrac{1}{4} + \cdots + \frac{1}{2^{N-1}}$$

$$< \tfrac{9}{4}$$

for all N. So $(s_{2^N})_{N\geq 1}$ is increasing bounded above, therefore converges. But therefore, also, the full sequence $(s_N)_{N\geq 1}$ must converge, since otherwise it would diverge to infinity, which would involve its subsequence $(s_{2^N})_{N\geq 1}$ also diverging to infinity. Hence $\sum_1^\infty 1/n^2$ converges. □

Observe that, even though we can show $\sum_1^\infty 1/n^2$ converges by the above method, we cannot say what its sum is. In fact, the sum is known to be $\pi^2/6$, a fact which was first discovered by Euler in 1734, but a proof of this result is beyond the scope of this book. (See Appendix.)

3.7 Exercise

Show that the series $\sum_1^\infty 1/n^3$ converges by the method of 3.6. □

The problem with series in general is that we cannot expect a simple formula for the Nth partial sum. This means that indirect methods have to be used to determine whether or not a series converges (as in 3.4 and 3.6). Such methods may give little or no information as to the actual value of the sum of a convergent series. For many purposes, however, it will be enough to know that a series does converge, without worrying too much about what its sum is.

In order to enable quick decisions to be made on convergence of series, a number of simple tests have been devised. These mainly involve reference to the nth term rather than the Nth partial sum, clearly a simplifying manoeuvre. The tests we shall describe, whilst not forming an exhaustive collection, should nevertheless enable the reader to cope with most series he is likely to encounter in real life.

3.8 Theorem

(i) If $\sum_1^\infty a_n$, $\sum_1^\infty b_n$ both converge, then so does $\sum_1^\infty (a_n + b_n)$, and

$$\sum_1^\infty (a_n + b_n) = \sum_1^\infty a_n + \sum_1^\infty b_n.$$

(ii) If $\sum_1^\infty a_n$ converges, and C is constant, then $\sum_1^\infty (Ca_n)$ converges, and its sum is

$$\sum_1^\infty (Ca_n) = C\sum_1^\infty a_n.$$

Proof (i) For all N we have

$$\sum_1^N (a_n + b_n) = \sum_1^N a_n + \sum_1^N b_n,$$

so, taking limits and applying 2.11, we deduce immediately that $\Sigma_1^\infty (a_n + b_n)$ is convergent, and its sum is given by

$$\sum_1^\infty (a_n + b_n) = \sum_1^\infty a_n + \sum_1^\infty b_n.$$

(ii) Follows similarly by taking limits in the formula

$$\sum_1^N (Ca_n) = C\sum_1^N a_n.$$

□

3.9 Applications

(i) $\sum_1^\infty \left(\frac{1}{2^n} + \frac{1}{n^2}\right) = 1 + \frac{\pi^2}{6}.$

(ii) The general 'geometric series'

$$a + ar + ar^2 + \cdots + ar^n + \cdots$$

having first term a, and common ratio r, converges if $|r| < 1$, and its sum in this case is $a/(1-r)$.

In fact, by 3.2, we have

$$\begin{aligned} a + ar + ar^2 + \cdots &= a + a\sum_1^\infty r^n \\ &= a + \frac{ar}{1-r} \\ &= \frac{a}{1-r} \end{aligned}$$

if $|r| < 1$.

□

3.10 Theorem

If $\Sigma_1^\infty a_n$ converges, then the sequence $(a_n)_{n\geqslant 1}$ of nth terms is null.

Proof Let $s_N = \Sigma_1^N a_n$, and suppose that $(s_N)_{N\geqslant 1}$ converges to s. Then we have

$$a_n = s_n - s_{n-1}$$

and so $(a_n)_{n\geqslant 1}$ must converge to $s - s = 0$.

□

3.11 Corollary: Non-null test

If (a_n) is *not* null, then $\sum_1^\infty a_n$ must diverge. □

3.12 Application

The non-null test can be used to give another proof of the divergence of $\sum_1^\infty x^n$ when $|x| \geqslant 1$. (See 3.2.) In fact, $(x^n)_{n \geqslant 1}$ is not null for $|x| \geqslant 1$ (see 2.26), so the result is immediate. □

Please note that the non-null test is a test for *divergence* only. One cannot prove a series $\sum_1^\infty a_n$ *converges* by demonstrating that its sequence of nth terms $(a_n)_{n \geqslant 1}$ is null. For example, $(1/n)_{n \geqslant 1}$ is null and yet $\sum_1^\infty 1/n$ diverges (see 3.4).

Faced with a series whose nth term is null, one has to try some other test, such as the following.

3.13 Comparison test

If $0 \leqslant a_n \leqslant b_n$ for all n, and if $\sum_1^\infty b_n$ converges, then also $\sum_1^\infty a_n$ converges.

Proof If we write $s_N = \sum_1^N a_n$, $t_N = \sum_1^N b_n$, then we have $s_N \leqslant t_N$ for all N, and we know $(t_N)_{N \geqslant 1}$ converges. It follows (by 2.29) that (t_N) is bounded above, and therefore so is (s_N). However, (s_N) is *increasing*, on account of the fact that $a_n \geqslant 0$ for all n. Hence (s_N) must converge (by 2.33), i.e., $\sum_1^\infty a_n$ converges. □

Observe that the condition $a_n \geqslant 0$ for all n is crucial. Without this condition we could not say (s_N) is monotonic, and the proof of 3.13 would fail.

3.14 Example

$$\sum_1^\infty \frac{n+5}{n^3+7} \text{ converges.}$$

In fact, we have

$$0 \leqslant \frac{n+5}{n^3+7} \leqslant \frac{6n}{n^3} = \frac{6}{n^2}$$

for all $n \geqslant 1$, and $\sum_1^\infty (6/n^2)$ converges by 3.4 and 3.8. □

3.15 Exercise

Show the following series converge.

$$\text{(i)} \sum_1^\infty \frac{n^2+2}{n^4+7} \qquad \text{(ii)} \sum_1^\infty \frac{2^n+3^n}{4^n+5^n}. \qquad \square$$

3.16 Corollary to 3.13

If $0 \leqslant a_n \leqslant b_n$ for all n, and if $\sum_1^\infty a_n$ diverges, then also $\sum_1^\infty b_n$ diverges.

Proof Follows immediately from 3.13. $\square$

3.17 Example

$$\sum_1^\infty \frac{n+5}{n^2+7} \text{ diverges.}$$

In fact, we have

$$\frac{n+5}{n^2+7} \geqslant \frac{n}{8n^2} = \frac{1}{8n}$$

for all $n \geqslant 1$, and $\sum_1^\infty (1/8n)$ diverges (by 3.2 and 3.8). $\square$

3.18 Exercises

Show the following series diverge.

$$\text{(i)} \sum_1^\infty \frac{n^2+2}{n^3+7} \qquad \text{(ii)} \sum_1^\infty \frac{6^n+5^n}{4^n+3^n}. \qquad \square$$

The following slightly more general form of the comparison test is sometimes useful.

3.19 Theorem

If $0 \leqslant a_n \leqslant b_n$ for all $n \geqslant$ some N, and if $\sum_1^\infty b_n$ converges, then so does $\sum_1^\infty a_n$.

Proof If $s_n = \sum_{r=1}^n a_r$, $t_n = \sum_{r=1}^n b_r$, then we have

$$s_n - s_N = \sum_{r=N+1}^n a_r \leqslant \sum_{r=N+1}^n b_r = t_n - t_N$$

for all $n > N$. Hence, as in the proof of 3.13, the boundedness of (t_n) implies the boundedness of (s_n), and (s_n) increases for $n \geqslant N$, so converges by 2.33. □

3.20 Example

$$\sum_1^{\infty} \left(\frac{1}{2} + \frac{1}{n}\right)^n \text{ converges.}$$

In fact, we have

$$\left(\frac{1}{2} + \frac{1}{n}\right)^n \leqslant \left(\frac{5}{6}\right)^n$$

for all $n \geqslant 3$, so the result follows by 3.2 and 3.19. □

Our next test is one of the easiest tests to apply in that it gives a direct method of determining convergence or divergence of many series by evaluation of a simple limit.

3.21 Ratio test

If $a_n > 0$ for all $n \geqslant 1$, and if the sequence $(a_{n+1}/a_n)_{n \geqslant 1}$ converges to a limit l, then the series $\sum_1^{\infty} a_n$ converges if $l < 1$, and diverges if $l > 1$.

Proof

Case l < 1 Let $r = \frac{1}{2}(l+1)$, so that $l < r < 1$. Then there must exist N such that

$a_{n+1}/a_n < r$ for all $n \geqslant N$. (Play $\varepsilon = r - l$.) Therefore, if $n > N$, we have

$$0 < a_n = \frac{a_n}{a_{n-1}} \frac{a_{n-1}}{a_{n-2}} \cdots \frac{a_{N+1}}{a_N} a_N < r^{n-N} a_N,$$

and $\sum_{n=1}^{\infty} r^{n-N} a_N$ converges, since it is a geometric series with common ratio r satisfying $0 < r < 1$. Hence $\sum_1^{\infty} a_n$ converges by 3.19.

Case l > 1 In this case there exists N such that $a_{n+1}/a_n > 1$ for all $n \geqslant N$. (Play $\varepsilon = l - 1$.) Therefore (a_n) increases (strictly) for $n \geqslant N$, and so cannot be null. Hence $\sum_1^{\infty} a_n$ diverges by the non-null test (3.11). □

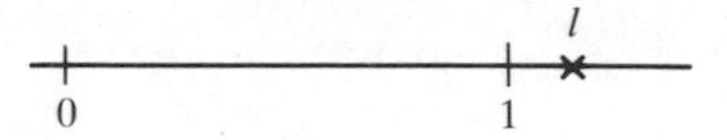

Observe that we say nothing about the case $l=1$. In fact, if $l=1$, $\Sigma_1^\infty a_n$ may either converge or diverge. For example, $\Sigma_1^\infty 1/n^2$ converges and $\Sigma_1^\infty 1/n$ diverges, and yet $l=1$ in both cases.

3.22 Example

$\Sigma_1^\infty n/2^n$ converges.

In fact, writing $a_n = n/2^n$, we have

$$\frac{a_{n+1}}{a_n} = \frac{n+1}{2n}$$

which converges to $\frac{1}{2}$, so we are in the case $l<1$. □

3.23 Exercises

Use the ratio test to discover which of the following series converge:

$$\text{(i) } \sum_1^\infty \frac{n^2}{2^n} \qquad \text{(ii) } \sum_1^\infty \frac{n!}{2^n}$$ □

The condition $a_n \geqslant 0$ for the comparison test and tests based on it, such as the ratio test (see also 3.38, questions 7, 8, 9), may seem unduly restrictive. In fact, these tests are still useful for series $\Sigma_1^\infty a_n$ which don't satisfy this condition. This is because we can always test the corresponding series $\Sigma_1^\infty |a_n|$, which does satisfy the condition, and then use the following theorem.

3.24 Theorem

If $\Sigma_1^\infty |a_n|$ converges, then so does $\Sigma_1^\infty a_n$. □

3.25 Definition

We say the series $\Sigma_1^\infty a_n$ converges *absolutely* (or is *absolutely* convergent) if the series of absolute values $\Sigma_1^\infty |a_n|$ converges.

For example, the alternating series

$$\sum_1^\infty \frac{(-1)^{n-1}}{n^2} = 1 - \frac{1}{2^2} + \frac{1}{3^2} - \frac{1}{4^2} + \cdots$$

converges absolutely, whilst

$$\sum_1^\infty \frac{(-1)^{n-1}}{n} = 1 - \frac{1}{2} + \frac{1}{3} - \frac{1}{4} + \cdots$$

doesn't.

3.24 says absolute convergence implies convergence. So, e.g. $\sum_1^\infty (-1)^{n-1}/n^2$ converges. Of course, 3.24 says nothing about the convergence of $\sum_1^\infty (-1)^{n-1}/n$, since this series is not absolutely convergent. In fact, we shall see shortly that $\sum_1^\infty (-1)^{n-1}/n$ *is* convergent, which shows that the converse of 3.24 is false.

Proof of 3.24

Let a_n^+, a_n^- be defined as follows.

$$\begin{aligned} a_n^+ &= a_n && \text{if } a_n \geqslant 0, \\ &= 0 && \text{otherwise.} \\ a_n^- &= |a_n| && \text{if } a_n \leqslant 0, \\ &= 0 && \text{otherwise.} \end{aligned}$$

For example, if

$$\sum_1^\infty a_n = 1 - \tfrac{1}{2} + \tfrac{1}{3} - \tfrac{1}{4} + \tfrac{1}{5} - \tfrac{1}{6} + \cdots$$

then

$$\sum_1^\infty a_n^+ = 1 + 0 + \tfrac{1}{3} + 0 + \tfrac{1}{5} + 0 + \cdots,$$

$$\sum_1^\infty a_n^- = 0 + \tfrac{1}{2} + 0 + \tfrac{1}{4} + 0 + \tfrac{1}{6} + \cdots.$$

Observe that $a_n^+ \geqslant 0$, $a_n^- \geqslant 0$, and

(i) $a_n^+ + a_n^- = |a_n|$,

(ii) $a_n^+ - a_n^- = a_n$.

It follows from (i) and the comparison test that $\sum_1^\infty a_n^+$, $\sum_1^\infty a_n^-$ both converge. Therefore, by (ii) and 3.8, $\sum_1^\infty a_n$ converges. □

For series which are not absolutely convergent, the question of convergence remains unresolved, and is in many cases not easy to resolve. However, there is one class of series for which convergence can be established without undue difficulty by means of the following test.

3.26 Alternating series test

If $(a_n)_{n\geqslant 1}$ is a decreasing null sequence, then the alternating series $\sum_1^\infty (-1)^{n-1} a_n$ converges. □

3.27 Example

$\sum_1^\infty (-1)^{n-1}/n$ converges.
In fact, $a_n = 1/n$ clearly satisfies the conditions of 3.26. □

Hence, as claimed earlier, Theorem 3.24 has no converse. We emphasize this fact by making the following definition.

3.28 Definition

The series $\sum_1^\infty a_n$ will be said to converge *conditionally* (or to be *conditionally* convergent) if it converges, but not absolutely. □

So, e.g., $\sum_1^\infty (-1)^{n-1}/n$ converges conditionally.

Proof of 3.26

If we write $s_N = \sum_1^N (-1)^{n-1} a_n$, then we have

$$\begin{aligned} s_{2N} &= (a_1 - a_2) + (a_3 - a_4) + \cdots + (a_{2N-1} - a_{2N}) \\ &\leqslant a_1, \end{aligned}$$

$0 \qquad a_4 \qquad a_3 \qquad a_2 \qquad a_1$

showing that $(s_{2N})_{N\geqslant 1}$ is increasing bounded above, therefore convergent to s, say. But then

$$s_{2N-1} = s_{2N} + a_{2N}$$

also converges to s. Hence the full sequence $(s_N)_{N\geqslant 1}$ converges to s.

3.29 Exercises

Which of the following series converge absolutely or conditionally?

$$\text{(i)}\ \sum_1^\infty (-1)^n \frac{n+1}{n^2+1} \qquad \text{(ii)}\ \sum_1^\infty (-1)^n \frac{n+1}{n^3+1}$$

N.B. Don't forget to check that a_n *decreases* when applying the alternating series test. □

We shall now illustrate the application of the tests so far mentioned by considering a certain class of series known as 'power series' defined as follows.

3.30 Definition

A *power series* is a series of the form $\sum_{n=0}^{\infty} a_n x^n$, where $(a_n)_{n\geq 0}$ is a sequence of constants, and x is a variable. □

Our interest in power series stems from the fact that Maclaurin series are power series, and that it is our intention to use such series to define some of the important functions of analysis in the next chapter.

3.31 Examples

(i) $\sum_0^\infty x^n$. This is the case $a_n = 1$ for all n. This power series is already known to be absolutely convergent if $|x| < 1$, and divergent otherwise. (See 3.2.)

(ii) $\sum_1^\infty x^n/n$. If we write $b_n = x^n/n$, then we have

$$\left|\frac{b_{n+1}}{b_n}\right| = \left|\frac{x^{n+1}}{n+1}\frac{n}{x^n}\right| = \frac{n\,|x|}{n+1},$$

which converges to $|x|$. Therefore, by the ratio test (3.21), this power series is absolutely convergent for $|x| < 1$, and divergent if $|x| > 1$ (since its sequence of nth terms is then not null, see proof of 3.21).

This leaves the cases $x = \pm 1$. If $x = 1$, the series is $\sum_1^\infty 1/n$, which we know diverges. If $x = -1$, the series is $\sum_1^\infty (-1)^n/n$, which converges by 3.27 and 3.8 (ii).

Hence $\sum_1^\infty x^n/n$ converges if $-1 \leq x < 1$, and diverges otherwise.

(iii) $\sum_1^\infty x^n/n^2$. Writing $b_n = x^n/n^2$, we have

$$\left|\frac{b_{n+1}}{b_n}\right| = \frac{n^2\,|x|}{(n+1)^2}$$

which again converges to $|x|$, showing the series to be absolutely convergent if $|x| < 1$, and divergent if $|x| > 1$. On this occasion, we have absolute convergence at $x = \pm 1$, since then $\sum_1^\infty |b_n| = \sum_1^\infty 1/n^2$.

Hence, $\sum_1^\infty x^n/n^2$ converges absolutely if $|x| \leq 1$, and diverges otherwise. □

3.32 Exercise

Discuss the convergence and absolute convergence of the following power series for various values of x.

$$\text{(i)}\ \sum_1^\infty \frac{x^n}{n!} \qquad \text{(ii)}\ \sum_1^\infty \frac{(-1)^n}{n} x^{2n}$$ □

The above examples demonstrate the various possibilities that can occur within the constraints of the following general theorem on the convergence of power series.

3.33 Theorem: Radius of convergence

For any power series $\sum_0^\infty a_n x^n$, precisely one of the following alternatives must occur.

(i) $\sum_0^\infty a_n x^n$ converges for all x.

(ii) $\sum_0^\infty a_n x^n$ converges only if $x = 0$.

(iii) There exists a real number $R > 0$, called the *radius of convergence*, such that $\sum_0^\infty a_n x^n$ converges if $|x| < R$ and diverges if $|x| > R$.

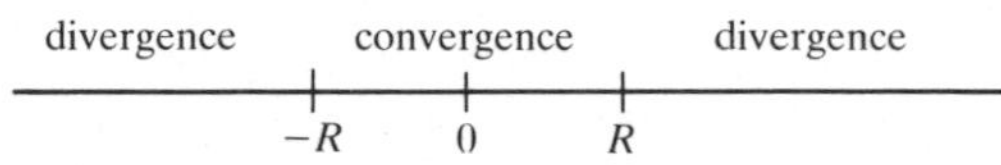

We describe case (i) by saying $R = \infty$, and case (ii) by saying $R = 0$. Examples of series in these categories are, respectively, $\sum_1^\infty x^n/n!$, $\sum_1^\infty n!\,x^n$. In case (iii), no general statement can be made about behaviour at $x = \pm R$. The examples 3.31 above all have $R = 1$, but exhibit different behaviour at $x = \pm 1$.

Proof of 3.33 Is achieved by means of the following.

Key lemma If $\sum_0^\infty a_n x^n$ converges when $x = X$, then it converges absolutely for all x satisfying $|x| < |X|$.

Proof of key lemma Observe that, if $|x| < |X|$, then

$$|a_n x^n| = |a_n X^n| \left|\frac{x}{X}\right|^n \leqslant M \left|\frac{x}{X}\right|^n,$$

where $M = \sup_{n \geqslant 0} |a_n X^n|$, which exists since $(a_n X^n)_{n \geqslant 0}$ is a null sequence (by 3.10) and is therefore bounded (by 2.29). Hence $\sum_0^\infty |a_n x^n|$ converges by comparison with $\sum_0^\infty |x/X|^n$, a geometric series with common ratio $|x/X| < 1$. □

Proof of 3.33 Suppose we are not in cases (i) or (ii). We shall show that we must then be in case (iii).

Let the set E be defined as follows.

$$E = \left\{x : \sum_0^\infty a_n x^n \text{ converges}\right\}.$$

The assumption that we are not in case (i) ensures that E is bounded since, if X is such that $\sum_0^\infty a_n X^n$ diverges, then, by the key lemma, $\sum_0^\infty a_n x^n$ cannot converge for any x satisfying $|x| > |X|$. The fact that any power series converges trivially at $x = 0$ ensures that E is non-empty. Therefore, by the upper bound axiom, E has a supremum.

Let $R = \sup E$. The assumption that we are not in case (ii) implies that $R > 0$, since, if $X \neq 0$ is such that $\sum_0^\infty a_n X^n$ converges, then, by the key lemma, $\sum_0^\infty a_n x^n$ also converges for all $|x| > |X|$, showing $R \geqslant |X|$.

Suppose $|x| < R$. Then there must exist X satisfying $|x| < X \leqslant R$ such that $\sum_0^\infty a_n X^n$ converges (since $R = \sup E$). Therefore, by the key lemma, $\sum_0^\infty a_n x^n$ converges (absolutely).

X

$-R$ 0 $|x|$ R

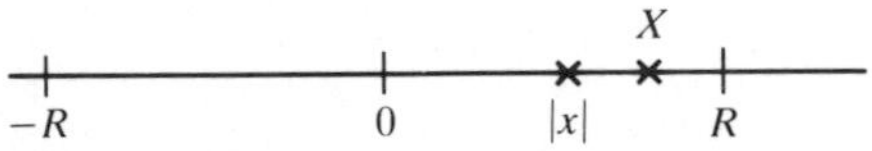

Suppose $|x| > R$. Then, by the key lemma, convergence of $\sum_0^\infty a_n x^n$ for this x would imply convergence of $\sum_0^\infty a_n X^n$, where

X

$-R$ 0 R $|x|$

$X = \frac{1}{2}(|x| + R) > R$, which would contradict the definition of R. Therefore $\sum_0^\infty a_n x^n$ must diverge. □

Observe that we have proved $\sum_0^\infty a_n x^n$ converges *absolutely* for $|x| < R$. Observe also that, when we are in the case $R = \infty$, we have *absolute* convergence of $\sum_0^\infty a_n x^n$ for all x (by the key lemma).

3.34 Exercises

Find the radius of convergence of the following power series.

$$\text{(i)} \sum_0^\infty (2^n + 3^n) x^n \qquad \text{(ii)} \sum_1^\infty \frac{(2n)!}{(n!)^2} x^n$$

Hint. Use the ratio test. □

The reader may have noticed a general lack of theorems about arithmetical operations on series. The only results in this direction we have so far mentioned are 3.8 (i) and (ii) concerning addition and constant multiples of series. We shall now attempt to extend this somewhat limited repertoire by considering the possibility of multiplying series.

Suppose we have two series $\sum_1^\infty a_n$, $\sum_1^\infty b_n$ and we try to multiply them together. If we assume an infinite distributive law, we ought to be able to say

$$\left(\sum_{m=1}^{\infty} a_m\right)\left(\sum_{n=1}^{\infty} b_n\right) = \sum_{m,n=1}^{\infty} a_m b_n.$$

However, the question immediately arises as to what order the terms $a_m b_n$ of the series $\sum_{m,n=1}^{\infty} a_m b_n$ should be taken in.

If we were to assume an infinite commutative law of addition, then the ordering of the terms would be immaterial. Unfortunately, no such law exists as the following example shows.

Consider the alternating harmonic series

$$\sum_{1}^{\infty} \frac{(-1)^{n-1}}{n} = 1 - \tfrac{1}{2} + \tfrac{1}{3} - \tfrac{1}{4} + \cdots,$$

which we know to be convergent by the alternating series test (3.26). Let its sum be s. Suppose we rearrange the terms taking two negative terms for each positive term. We get

$$\begin{aligned} 1 - \tfrac{1}{2} - \tfrac{1}{4} + \tfrac{1}{3} - \tfrac{1}{6} - \tfrac{1}{8} + \cdots &= (1 - \tfrac{1}{2}) - \tfrac{1}{4} + (\tfrac{1}{3} - \tfrac{1}{6}) - \tfrac{1}{8} + \cdots \\ &= \tfrac{1}{2} - \tfrac{1}{4} + \tfrac{1}{6} - \tfrac{1}{8} + \cdots \\ &= \tfrac{1}{2}(1 - \tfrac{1}{2} + \tfrac{1}{3} - \tfrac{1}{4} + \cdots) \\ &= \tfrac{1}{2}s. \end{aligned}$$

Observe that reordering the terms has changed the sum! Perhaps it is not so strange after all that theorems on arithmetical manipulation of series appear to be objects of great rarity.

Whilst the situation is bleak in general, there is a class of series for which rearrangement and multiplication can be done with impunity, namely the *absolutely* convergent series. Observe that the counter-example above involves a *conditionally* convergent series. Indeed the term 'conditionally' convergent originates in the fact that the sum of such series, indeed their very convergence, is conditional on the order in which their terms are taken.

3.35 Theorem

If $\sum_1^{\infty} a_n$ is absolutely convergent, and if $\sum_1^{\infty} b_n$ is a rearrangement, i.e. the same terms but taken in a different order, then $\sum_1^{\infty} b_n$ is also absolutely convergent and has the same sum. □

Proof We assume first that $a_n \geqslant 0$ for all n. For each N, we can choose N'

such that $b_1, b_2, \ldots, b_N$ all occur among $a_1, a_2, \ldots, a_{N'}$. Therefore

$$\sum_1^N b_n \leqslant \sum_1^{N'} a_n \leqslant \sum_1^\infty a_n.$$

Hence $\sum_1^N b_n$ is bounded above, and so converges. Also

$$\sum_1^\infty b_n \leqslant \sum_1^\infty a_n. \qquad (2.33)$$

However, reversing the roles of a_n, b_n gives similarly

$$\sum_1^\infty a_n \leqslant \sum_1^\infty b_n.$$

Hence

$$\sum_1^\infty a_n = \sum_1^\infty b_n.$$

We now drop the assumption that $a_n \geqslant 0$. As in the proof of 3.24, let

$$\begin{aligned} a_n^+ &= a_n && \text{if } a_n \geqslant 0, \\ &= 0 && \text{otherwise}, \\ a_n^- &= |a_n| && \text{if } a_n \leqslant 0, \\ &= 0 && \text{otherwise}, \end{aligned}$$

and let b_n^+, b_n^- be defined similarly. Then $\sum_1^\infty b_n^+$, $\sum_1^\infty b_n^-$ are the corresponding rearrangements of $\sum_1^\infty a_n^+$, $\sum_1^\infty a_n^-$, which are both convergent series of positive terms (since $\sum_1^\infty a_n$ is absolutely convergent). Therefore, by what we have already proved, $\sum_1^\infty b_n^+$, $\sum_1^\infty b_n^-$ both converge and $\sum_1^\infty b_n^+ = \sum_1^\infty a_n^+$, $\sum_1^\infty b_n^- = \sum_1^\infty a_n^-$. Hence $\sum_1^\infty b_n$ is absolutely convergent, since

$$|b_n| = b_n^+ + b_n^-,$$

and its sum is given by

$$\begin{aligned} \sum_1^\infty b_n &= \sum_1^\infty b_n^+ - \sum_1^\infty b_n^- \\ &= \sum_1^\infty a_n^+ - \sum_1^\infty a_n^- \\ &= \sum_1^\infty a_n. \end{aligned}$$

3.36 Theorem

If $\sum_{m=1}^{\infty} a_m$, $\sum_{n=1}^{\infty} b_n$ are absolutely convergent, and if $\sum_{p=1}^{\infty} c_p$ is the series obtained by taking the products $a_m b_n$ is any order, then $\sum_{p=1}^{\infty} c_p$ is absolutely convergent and

$$\sum_{p=1}^{\infty} c_p = \left(\sum_{m=1}^{\infty} a_m\right)\left(\sum_{n=1}^{\infty} b_n\right).$$

Proof Again, we assume firstly that $a_m \geqslant 0$, $b_n \geqslant 0$. Let the products $a_m b_n$ be written in an infinite matrix as shown below.

$$\begin{matrix} a_1b_1 & a_1b_2 & a_1b_3 \cdots \\ a_2b_1 & a_2b_2 & a_2b_3 \cdots \\ a_3b_1 & a_3b_2 & a_3b_3 \cdots \\ \vdots & \vdots & \vdots \end{matrix}$$

And let, in the first instance, $\sum_{p=1}^{\infty} c_p$ represent $\sum_{m,n=1}^{\infty} a_m b_n$ taken in the following order.

$$\begin{matrix} c_1 & c_4 & c_9 \cdots \\ c_2 & c_3 & c_8 \cdots \\ c_5 & c_6 & c_7 \cdots \\ \vdots & \vdots & \vdots \end{matrix}$$

Then, for each N, we have

$$\sum_{p=1}^{N^2} c_p = \left(\sum_{m=1}^{N} a_m\right)\left(\sum_{n=1}^{N} b_n\right).$$

Taking limits, this shows that a subsequence of the partial sums of $\sum_{p=1}^{\infty} c_p$ converges to $(\sum_{m=1}^{\infty} a_m)(\sum_{n=1}^{\infty} b_n)$. Therefore, since $c_p \geqslant 0$, the whole sequence does, i.e.

$$\sum_{p=1}^{\infty} c_p = \left(\sum_{m=1}^{\infty} a_m\right)\left(\sum_{n=1}^{\infty} b_n\right).$$

By 3.35 the same result holds for any rearrangement of $\sum_{p=1}^{\infty} c_p$.

Suppose now that we drop the assumption that $a_m \geqslant 0$, $b_n \geqslant 0$. The above argument applied to $\sum_m |a_m|$, $\sum_n |b_n|$, $\sum_p |c_p|$ shows that $\sum_p c_p$ is absolutely convergent. Therefore $\sum_p c_p$ is convergent, and its sum is independent of the order in which its terms are taken. If we adopt the special ordering indicated at the beginning of the proof, we still have

$$\sum_{p=1}^{N^2} c_p = \left(\sum_{m=1}^{N} a_m\right)\left(\sum_{n=1}^{N} b_n\right),$$

which, on taking limits, gives

$$\sum_{p=1}^{\infty} c_p = \left(\sum_{m=1}^{\infty} a_m\right)\left(\sum_{n=1}^{\infty} b_n\right).$$

□

3.37 Application: Multiplying power series

Suppose we have two power series $\sum_0^\infty a_n x^n$, $\sum_0^\infty b_n x^n$. If we formally multiply them, we obtain a third power series $\sum_0^\infty c_n x^n$, where

$$c_n = a_0 b_n + a_1 b_{n-1} + \cdots + a_n b_0.$$

This power series $\sum_0^\infty c_n x^n$ is called the *Cauchy product* of $\sum_0^\infty a_n x^n$, $\sum_0^\infty b_n x^n$. If $\sum_0^\infty a_n x^n$, $\sum_0^\infty b_n x^n$ both have radius of convergence R, then they converge absolutely for all $|x| < R$ (see 3.33); therefore, by 3.36, $\sum_0^\infty c_n x^n$ must have radius of convergence at least R, and

$$\sum_0^\infty c_n x^n = \left(\sum_0^\infty a_n x^n\right)\left(\sum_0^\infty b_n x^n\right)$$

for all $|x| < R$.

For example, the Cauchy product of $\sum_0^\infty x^n$ with itself is $\sum_0^\infty (n+1)x^n$, so we have

$$\sum_0^\infty (n+1)x^n = \frac{1}{(1-x)^2}$$

for all $|x| < 1$. It is interesting to observe that this result can be obtained in two other ways. One can either differentiate the formula

$$\sum_0^\infty x^n = \frac{1}{1-x},$$

or one can expand $(1-x)^{-2}$ by the binomial theorem. Neither of these alternative approaches can be justified rigorously at this stage of course.

□

3.38 Miscellaneous exercises

1. Which of the following series converge?

(i) $\sum_1^\infty \frac{1}{n^2+1}$ (ii) $\sum_1^\infty \frac{1}{n-\frac{1}{2}}$

(iii) $\sum_{1}^{\infty} \frac{1+2^n}{1+3^n}$ (iv) $\sum_{1}^{\infty} \frac{1}{3^n - 2^n}$

(v) $\sum_{1}^{\infty} \left(\frac{9}{10} + \frac{1}{n}\right)^n$ (vi) $\sum_{1}^{\infty} (\sqrt{(n+1)} - \sqrt{n})$

2. Which of the following series converge absolutely or conditionally?

(i) $\sum_{1}^{\infty} (-1)^n \frac{1}{n^2+1}$ (ii) $\sum_{1}^{\infty} (-1)^n \frac{n}{n^2+1}$

(iii) $\sum_{1}^{\infty} (-1)^n \frac{n^2}{n^2+1}$ (iv) $\sum_{1}^{\infty} (-1)^n (\sqrt{(n+1)} - \sqrt{n})$

(v) $\sum_{2}^{\infty} \frac{(-1)^n}{n + (-1)^n}$ (vi) $\sum_{2}^{\infty} \frac{(-1)^n}{n^2 + (-1)^n}$

3. Find the radius of convergence of the following power series.

(i) $\sum_{0}^{\infty} \frac{1+2^n}{1+3^n} x^n$ (ii) $\sum_{0}^{\infty} 2^{\sqrt{n}} x^n$

(iii) $\sum_{0}^{\infty} (\sqrt{(n+1)} - \sqrt{n}) x^n$ (iv) $\sum_{1}^{\infty} \frac{(3n)!}{n!\,(2n)!} x^n$

(v) $\sum_{1}^{\infty} \frac{n^n}{n!} x^n$ (vi) $\sum_{1}^{\infty} \left(\frac{1}{2} + \frac{1}{n}\right)^n x^n$

4. Show that, if for each $n \geqslant 1$, d_n is an integer between 0 and 9 inclusive, then the series $\sum_1^\infty d_n/10^n$ converges, and its sum satisfies

$$0 \leqslant \sum_{1}^{\infty} \frac{d_n}{10^n} \leqslant 1.$$

Show conversely that, given any real number x satisfying $0 \leqslant x \leqslant 1$, there exists a sequence $(d_n)_{n \geqslant 1}$ as above such that $\sum_1^\infty d_n/10^n = x$.

5. Prove that if α is a positive integer, and if $0 < x < 1$ is real, then the sequence $(n^\alpha x^n)_{n \geqslant 1}$ is null. *Hint*: Show $\sum_1^\infty n^\alpha x^n$ converges by the ratio test. Find the limit of the sequence

$$\left(\frac{2^n + n^3}{3^n + n^2}\right)_{n \geqslant 1}.$$

Find the radius of convergence of the power series

$$\sum_{1}^{\infty} \frac{x^n}{2^n - n}.$$

6. Discuss the convergence and absolute convergence of the series

$$1 - \frac{1}{2} + \frac{1}{\sqrt{3}} - \frac{1}{4} + \frac{1}{\sqrt{5}} - \frac{1}{6} + \cdots$$

N.B. Neither the ratio test nor the alternating series test is applicable here.

7. (Condensation test) Show that, if $0 \leqslant a_{n+1} \leqslant a_n$ for all n, then the series $\sum_1^\infty a_n$, $\sum_1^\infty 2^n a_{2^n}$ either both converge or both diverge.

(This is a refinement of the method of 3.4 and 3.6.)

Discuss the convergence of $\sum_1^\infty 1/n^\alpha$ for various values of α.

8. (nth root test) Show that, if $a_n \geqslant 0$, and if $\sqrt[n]{a_n}$ converges to a limit l, then $\sum_1^\infty a_n$ converges if $l < 1$, and diverges if $l > 1$. *Hint*: Compare with a geometric series. (cf. Proof of the ratio test (3.21).)

Use the nth root test to confirm your answer to question 3, part (vi).

What can you say if $l = 1$?

9. Show that, if $a_n > 0$, $b_n > 0$ and

$$\frac{a_{n+1}}{a_n} \leqslant \frac{b_{n+1}}{b_n}$$

for all $n \geqslant 1$, and if $\sum_1^\infty b_n$ converges, then also $\sum_1^\infty a_n$ converges.

Investigate the convergence of

$$\sum_{1}^{\infty} \frac{(2n)!}{(n!)^2} x^n$$

at $x = \pm\frac{1}{4}$.

10. Given that

$$1 - \tfrac{1}{2} + \tfrac{1}{3} - \tfrac{1}{4} + \cdots = s,$$

show that

(i) $1 + \frac{1}{3} - \frac{1}{2} + \frac{1}{5} + \frac{1}{7} - \frac{1}{4} + \cdots = \frac{3}{2}s$,

(ii) $1 - \frac{1}{2} + \frac{1}{3} + \frac{1}{5} - \frac{1}{4} + \frac{1}{7} + \frac{1}{9} + \frac{1}{11} - \frac{1}{6} + \cdots$ diverges to infinity.

Hints: (i) Observe that e.g.

$$\begin{aligned}(1 - \tfrac{1}{2} + \tfrac{1}{3} - \tfrac{1}{4} + \tfrac{1}{5} - \tfrac{1}{6} + \tfrac{1}{7} - \tfrac{1}{8}) &+ (\tfrac{1}{2} - \tfrac{1}{4} + \tfrac{1}{6} - \tfrac{1}{8}) \\ &= 1 + \tfrac{1}{3} - \tfrac{1}{2} + \tfrac{1}{5} + \tfrac{1}{7} - \tfrac{1}{4}.\end{aligned}$$

(ii) Use the A.M./H.M. inequality to show that the nth block of positive terms

$$\frac{1}{n^2-n+1}+\cdots+\frac{1}{n^2+n-1}>\frac{1}{n}.$$

11. Show that the Cauchy product of $\sum_1^\infty x^n/n$ with itself is

$$\sum_1^\infty \frac{2}{n+1}\left(1+\tfrac{1}{2}+\cdots+\frac{1}{n}\right)x^{n+1}.$$

Show that this series is conditionally convergent at $x=-1$, and deduce that its radius of convergence is 1. *Hint*: Let

$$a_n=\frac{1}{n+1}\left(1+\tfrac{1}{2}+\cdots+\frac{1}{n}\right).$$

Show that $a_n \geqslant a_{n+1}$ and

$$a_{2^n}<\frac{n+\frac{1}{2}}{2^n+1}$$

by the method of 3.6. Deduce (a_n) is null and use the alternating series test.

12. Show that the Cauchy product of $\sum_1^\infty x^n/\sqrt{n}$ with itself is $\sum_1^\infty c_n x^{n+1}$, where

$$c_n=\frac{1}{\sqrt{n}}+\frac{1}{\sqrt{[2(n-1)]}}+\frac{1}{\sqrt{[3(n-2)]}}+\cdots+\frac{1}{\sqrt{n}}.$$

Show this series diverges at $x=1$, and deduce that its radius of convergence is 1. *Hint*: Show that $c_n \geqslant 2$ for all n.

13. Prove that, if $\sum_1^\infty a_n$ converges, and if $\sum_1^\infty b_n$ is obtained from $\sum_1^\infty a_n$ by altering a finite number of terms, then $\sum_1^\infty b_n$ also converges, but its sum may differ from that of $\sum_1^\infty a_n$.

Hence give an alternative proof of 3.19.

14. Prove that, if $\sum_1^\infty a_n$ converges, and if $\sum_1^\infty b_n$ is obtained from $\sum_1^\infty a_n$ by grouping the terms, e.g.,

$$\sum_1^\infty b_n=(a_1+a_2)+(a_3+a_4+a_5+a_6)+(a_7)+\cdots$$

$$=b_1+b_2+b_3+\cdots,$$

then $\sum_1^\infty b_n$ converges, and $\sum_1^\infty b_n=\sum_1^\infty a_n$. *Hint*: Use 2.37.

15. Given that $\sum_1^\infty a_n$ converges absolutely, prove that

$$\left|\sum_1^\infty a_n\right| \leqslant \sum_1^\infty |a_n|.$$

Hint: Consider the proof of 3.24.

16. Given that $\sum_1^\infty a_n$ converges, prove that
   (i) for any $N \geqslant 1$, the series $\sum_{N+1}^\infty a_n$ converges,
   (ii) the sequence $(\sum_{N+1}^\infty a_n)_{N \geqslant 1}$ is null.

17. Show that if the power series $\sum_0^\infty a_n x^n$ has radius of convergence R, then the formally differentiated series $\sum_0^\infty n a_n x^{n-1}$ also has radius of convergence R.

4
Continuous functions

Having discussed convergence of sequences and series in Chapters 2 and 3, we now move to convergence of functions of a continuous variable. We shall need a theory of continuous convergence to give an adequate description of the process of differentiating a function (see Chapter 5). A theory of continuous convergence also enables a proper definition of the concept of a continuous function to be made. Since properties of continuous functions underlie important theorems in differential and integral calculus, it seems opportune to devote a chapter to continuous functions before proceeding to calculus.

A sequence can be thought of as a function of an integer variable n, and the problem of convergence is concerned with what happens as n gets large or 'tends to infinity'. With a function $f(x)$ of a continuous variable x, the question is what happens as x 'tends to' a particular value c. If $f(x)$ approaches a definite value l as x approaches c, we say $f(x)$ *tends to* l as x *tends to* c, and we write $f(x) \to l$ as $x \to c$. We call l the *limit* of $f(x)$ as x tends to c, and write $l = \lim_{x \to c} f(x)$. For example, $x^2 \to 4$ as $x \to 2$, equivalently, $\lim_{x \to 2} x^2 = 4$.

The arrow notation can also be used in the context of sequences and series. If the sequence $(a_n)_{n \geq 1}$ converges to a, we write $a_n \to a$ as $n \to \infty$, also $\lim_{n \to \infty} a_n = a$. If the series $\sum_1^\infty a_n$ converges, we can write $\sum_1^\infty a_n = \lim_{N \to \infty} \sum_1^N a_n$.

If we try to analyse what we mean by saying $f(x) \to l$ as $x \to c$, we find ourselves again in a two-handed game situation, just as we did for sequential convergence. Here we must be able to say that $f(x)$ is as near to l as *someone else* might want, provided we take x near enough to c. The rigorous definition traditionally makes use of the Greek letters ε, δ and is as follows (see Fig. 4.1).

4.1 Definition

We say $f(x) \to l$ as $x \to c$ if, for any given $\varepsilon > 0$, there exists $\delta > 0$ such that

$$|f(x) - l| < \varepsilon$$

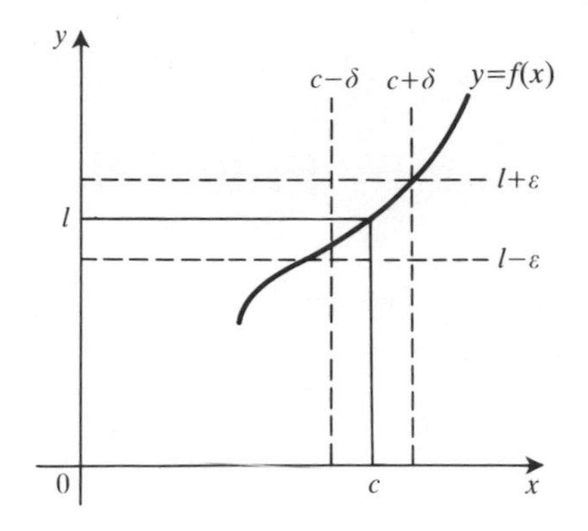

Fig. 4.1

for all x satisfying

$$0 < |x - c| < \delta. \qquad \square$$

Observe that, for the purposes of the present game, we require an ε-player and a δ-player. As usual the ε-player goes first, and the limit exists if the δ-player can come up with a strategy which copes with any play the ε-player may make.

Observe also that we are not interested in what happens when $x = c$. We want to know what happens as x *tends* to c without actually ever getting there.

4.2 Example

$x^2 \to 4$ as $x \to 2$ (see Fig. 4.2).

Suppose $\varepsilon > 0$ is given. We have to indicate how $\delta > 0$ can be

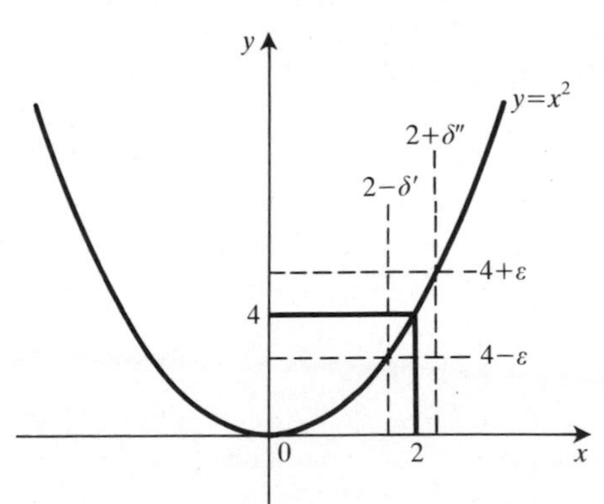

Fig. 4.2

chosen such that

$$|x^2 - 4| < \varepsilon$$

whenever

$$0 < |x - 2| < \delta.$$

Let $2 - \delta' = \sqrt{(4 - \varepsilon)}$, $2 + \delta'' = \sqrt{(4 + \varepsilon)}$ and take $\delta = \min\{\delta', \delta''\}$. Then $\delta > 0$ and, for any x satisfying

$$0 < |x - 2| < \delta,$$

we clearly have

$$|x^2 - 4| < \varepsilon. \qquad \square$$

A theory of limits for functions can now be built up in an analogous fashion to that for sequences. We can either translate the methods to the new situation, or we can use results from the sequential theory. The tie-up is provided by the following theorem.

4.3 Theorem

$f(x) \to l$ as $x \to c$ if and only if, for any sequence $(x_n)_{n \geq 1}$ such that $x_n \neq c$ for all n and $x_n \to c$ as $n \to \infty$, we have $f(x_n) \to l$ as $n \to \infty$.

Proof Suppose firstly that $f(x) \to l$ as $x \to c$, and that we have a sequence $x_n \neq c$ which $\to c$ as $n \to \infty$. We have to show $f(x_n) \to l$ as $n \to \infty$, so we must allow ourselves to be given $\varepsilon > 0$, and then indicate how we might choose a positive integer N such that

$$|f(x_n) - l| < \varepsilon$$

for all $n > N$. The fact that $f(x) \to l$ as $x \to c$ means we can choose $\delta > 0$ such that

$$|f(x) - l| < \varepsilon$$

whenever

$$0 < |x - c| < \delta$$

(by 4.1). And the fact that $x_n \to c$ as $n \to \infty$, and $x_n \neq c$ for all n means we can choose N such that

$$0 < |x_n - c| < \delta$$

for all $n > N$. Hence N has the required properties.

Now suppose conversely that we are told that $f(x_n) \to l$ whenever $(x_n)_{n \geq 1}$ is a sequence which converges to c but is never equal to c. We want to show this implies $f(x) \to l$ as $x \to c$. We shall achieve this by showing that, if $f(x)$ *doesn't* $\to l$ as $x \to c$, then we can construct a sequence $x_n \neq c$ converging to c such that $f(x_n)$ doesn't converge to $f(l)$. Failure of $f(x)$ to tend to l means that the ε-player can find ε for which the δ-player has no suitable δ. Therefore, for every positive integer n, playing $\delta = 1/n$ is inadequate to cope with this ε, so there must exist a real number, which we can call x_n, such that

$$0 < |x_n - c| < \frac{1}{n}$$

but

$$|f(x_n) - l| \geqslant \varepsilon.$$

This constructs a sequence (x_n) with the required properties. □

4.4 Application

We can use 4.3 to give an alternative proof of 4.2.

In fact, 2.11 gives immediately that, for any sequence $x_n \neq 2$ which $\to 2$, we have $x_n^2 \to 4$.

4.5 Theorem: Arithmetic of limits

If $f(x) \to l$ and $g(x) \to m$ as $x \to c$, then
(i) $f(x) + g(x) \to l + m$,
(ii) $f(g)g(x) \to lm$,
(iii) $f(x)/g(x) \to l/m$,
as $x \to c$, provided (for (iii)) $m \neq 0$ and $g(x) \neq 0$ for all $x \neq c$.

Proof (i) If $(x_n)_{n \geqslant 1}$ is any sequence such that $x_n \to c$ and $x_n \neq c$, then $f(x_n) \to l$ and $g(x_n) \to m$ by 4.3, and therefore, by 2.11, $f(x_n) + g(x_n) \to l + m$.

(ii) and (iii) are proved similarly. □

4.6 Corollaries

(i) If $f(x) \to l$ as $x \to c$, and C is a constant, then $Cf(x) \to Cl$ as $x \to c$.

(ii) If $f(x) \to l$ and $g(x) \to m$ as $x \to c$, then

$$f(x) - g(x) \to l - m$$

as $x \to c$. □

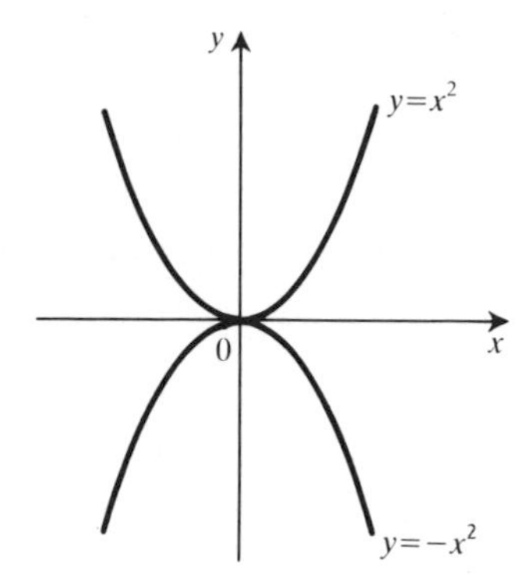

Fig. 4.3

4.7 Theorem: Taking limits in inequalities

If $f(x) \to l$, $g(x) \to m$ as $x \to c$, and if $f(x) \leqslant g(x)$ for all $x \neq c$, then $l \leqslant m$.

Proof Use 2.14 and 4.3. □

Just as for sequences, one cannot assume that a *strict* inequality persists in the limit. For example, $x^2 > -x^2$ for all $x \neq 0$, but $\pm x^2$ both $\to 0$ as $x \to 0$ (Fig. 4.3).

4.8 Theorem: Sandwich principle

If $f(x) \leqslant g(x) \leqslant h(x)$ for all $x \neq c$, and if $f(x)$, $h(x)$ both $\to l$ as $x \to c$, then also $g(x) \to l$ as $x \to c$.

Proof Follows from the sandwich principle for sequences (2.16) and 4.3. □

It is sometimes necessary to consider one-sided limits. For example, $f(x) = \sqrt{x}$ can only be defined for $x \geqslant 0$, so it would only be possible to consider the limit as $x \to 0$ through *positive* values. Accordingly, we make the following definition.

4.9 Definition

We say $f(x)$ tends to l as x tends to c *from the right*, and we write $f(x) \to l$ as $x \to c_+$ (also $l = \lim_{x \to c_+} f(x)$) if, for any given $\varepsilon > 0$, there exists $\delta > 0$ such that

$$|f(x) - l| < \varepsilon$$

for all x satisfying

$$c < x < c + \delta.$$

4.10 Example

$\sqrt{x} \to 0$ as $x \to 0_+$.

In fact, if $\varepsilon > 0$ is given, then, for $\delta = \varepsilon^2$, we have $\sqrt{x} < \varepsilon$ whenever $0 < x < \delta$. □

We define limits on the left, with the notation $x \to c_-$, similarly.

Clearly $\lim_{x \to c} f(x) = l$ if and only if $\lim_{x \to c_+} f(x) = \lim_{x \to c_-} f(x) = l$. It is conceivable that the limits from the left and right may differ. Consider the following example.

4.11 Example: The sign function

We define the *sign* function, which we abbreviate to $\operatorname{sgn} x$, as follows (see Fig. 4.4):

$$\begin{aligned} \operatorname{sgn} x &= 1 \quad \text{if } x > 0, \\ &= 0 \quad \text{if } x = 0, \\ &= -1 \quad \text{if } x < 0. \end{aligned}$$

If $f(x) = \operatorname{sgn} x$, we have $\lim_{x \to 0_+} f(x) = 1$, $\lim_{x \to 0_-} f(x) = -1$. It follows that $\lim_{x \to 0} f(x)$ cannot exist in this case. □

This is our first instance of non-existence of a limit. Other ways in which a limit may fail to exist are illustrated in the following examples.

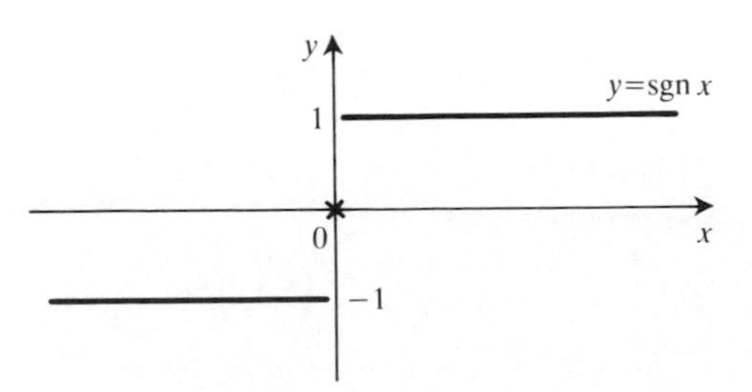

Fig. 4.4

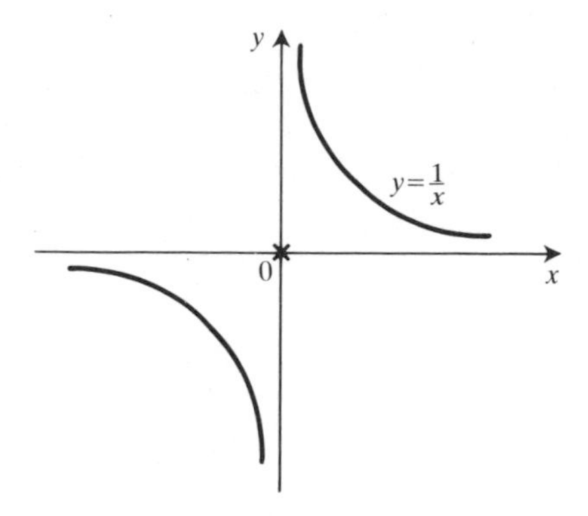

Fig. 4.5

4.12 Example

Let $f(x)$ be defined thus (see Fig. 4.5):

$$f(x) = \frac{1}{x} \quad \text{if } x \neq 0,$$
$$= 0 \quad \text{if } x = 0.$$

Observe that the formula $1/x$ doesn't define a value at $x = 0$. If we want our function $f(x)$ to have a value at $x = 0$, we must make a special definition for this point. We have chosen to define $f(0) = 0$, though we could have taken any other value if we had liked.

$\lim_{x\to 0} f(x)$ doesn't exist for this function since, e.g., if $x_n = 1/n$, then the sequence $f(x_n)$ diverges to infinity. Also, e.g., for $x_n = -1/n$, the sequence $f(x_n)$ diverges to minus infinity. □

The above example leads us to make the following definition.

4.13 Definition

We say $f(x)$ *tends to infinity* as x tends to c from the right, and we write $f(x) \to \infty$ as $x \to c_+$ (also $\lim_{x\to c_+} f(x) = \infty$) if, given any $C > 0$, there exists $\delta > 0$ such that $f(x) > C$ for all x satisfying $c < x < c + \delta$. □

Observe that the two-handed game here involves a C-player and a δ-player. Clearly $1/x \to \infty$ as $x \to 0_+$. Clearly also, there is an analogue of 4.3 which says that, in general, $f(x) \to \infty$ as $x \to c_+$ if and only if, for any sequence $x_n > c$ which converges to c, the sequence $f(x_n)$ diverges to infinity.

One can similarly define $f(x) \to \infty$ as $x \to c_-$, or as $x \to c$, also $f(x) \to -\infty$ as $x \to c$ from left or right or both sides simultaneously. For example, $1/x \to -\infty$ as $x \to 0_-$. □

It is important to emphasize that we do *not* regard ∞ as a number. When we write $\lim_{x \to c} f(x) = \infty$, we mean $f(x)$ *diverges* to infinity as $x \to c$. The usage $\lim_{x \to c} f(x) = \infty$ is of course an abuse of notation, since in this situation $f(x)$ has *no limit* as $x \to c$. It is nevertheless a convenient usage in many contexts and, properly understood, should not lead to any confusion.

Another example of non-existence of a limit is the following.

4.14 Example

Let $f(x)$ (see Fig. 4.6) be defined by saying

$$f(x) = \sin\frac{1}{x} \qquad \text{if } x \neq 0,$$
$$= 0 \qquad \text{if } x = 0.$$

$f(x)$ has no limit as $x \to 0_+$ since, if we let $x_n = 1/(n + \frac{1}{2})\pi$, then we have $f(x_n) = (-1)^n$, which is an oscillating sequence. Obviously $\lim_{x \to 0_-} f(x)$ doesn't exist either for similar reasons. □

Please note that officially we have not yet defined the function $\sin x$. We shall nevertheless take the liberty of using $\sin x$ in examples and illustrations on the understanding that it will in due course be given a rigorous definition and its properties rigorously established. (See 4.45 *et seq.*)

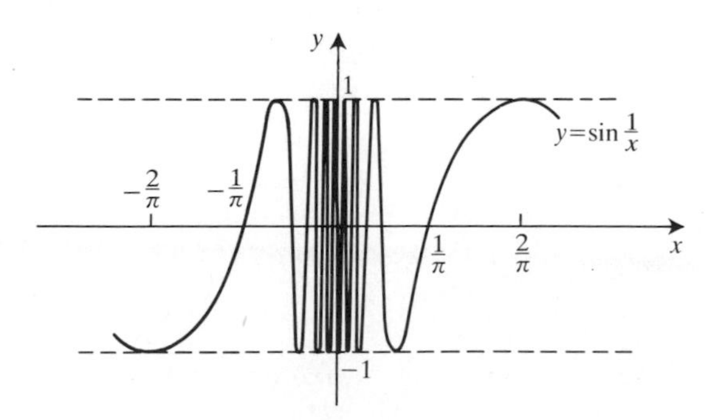

Fig. 4.6

One more type of limit remains to be considered, and that is where the *variable* diverges to infinity. The following two definitions are typical.

4.15 Definitions

(i) We say $f(x)$ tends to l as x tends to infinity, and we write $f(x) \to l$ as $x \to \infty$, or $\lim_{x\to\infty} f(x) = l$, if, given any $\varepsilon > 0$, there exists $C > 0$ such that

$$|f(x) - l| < \varepsilon$$

for all $x > C$.

(ii) We say $f(x)$ tends to infinity as x tends to infinity, and we write $f(x) \to \infty$ as $x \to \infty$, or $\lim_{x\to\infty} f(x) = \infty$, if, given any $C > 0$, there exists $C' > 0$ such that $f(x) > C$ for all $x > C'$. □

Definitions for $x \to -\infty$ are similar.

4.16 Examples

(i) $f(x) = 1/(1 + x^2)$ (Fig. 4.7). Clearly $f(x) \to 0$ as $x \to \pm\infty$ here.

(ii) $f(x) = x^3$ (Fig. 4.8). Clearly $f(x) \to \pm\infty$ as $x \to \pm\infty$ in this case. □

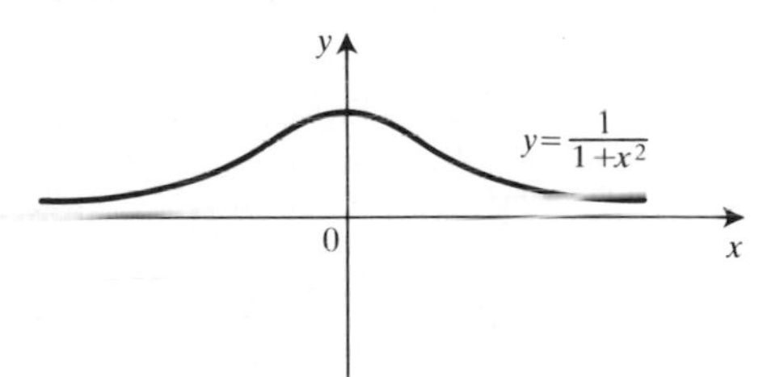

Fig. 4.7

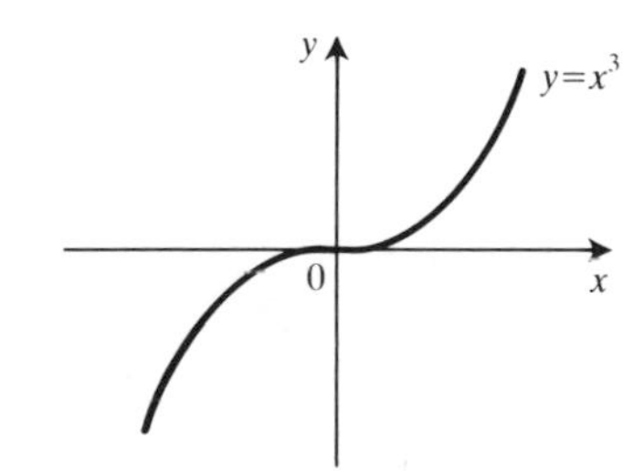

Fig. 4.8

The analogue of 4.3 for limits as $x \to \infty$ is that $f(x) \to l$ as $x \to \infty$ if and only if $f(x_n) \to l$ as $n \to \infty$ for any sequence x_n which $\to \infty$ as $n \to \infty$. Theorems 4.5 to 4.8 are all true for limits as $x \to \infty$ with the same proofs. These remarks also apply to limits as $x \to -\infty$.

4.17 Exercises

Investigate the existence or otherwise of the following limits.

(i) $\lim_{x\to 0} |\operatorname{sgn} x|$ (ii) $\lim_{x\to 0} x \sin \frac{1}{x}$ (iii) $\lim_{x\to 0} \frac{1}{x} \sin \frac{1}{x}$

Draw graphs in each case. □

Such is the theory of continuous limits. We now turn to continuous functions.

By a continuous function we mean one which has a continuous graph, i.e., the graph is a single unbroken curve which can be drawn without lifting pen from paper, as in Fig. 4.9.

Among the functions we have considered so far, it is clear that x^2, x^3, $1/(1+x^2)$ are continuous, but that $\operatorname{sgn} x$, $1/x$, $\sin 1/x$ are not.

It is the discontinuous functions which provide the clue as to how the property of being continuous should be rigorously defined. For example, the graph of $\operatorname{sgn} x$ has a break at $x = 0$, and this is because $\lim_{x\to 0+} \operatorname{sgn} x = 1$, $\lim_{x\to 0-} \operatorname{sgn} x = -1$ but $\operatorname{sgn} 0 = 0$. The functions $1/x$, $\sin 1/x$ have discontinuities at $x = 0$ because they have no limits there. This suggests that the condition for continuity should be that the function should converge to its value at every point. Discontinuous functions are those which converge to the wrong value, or converge to no value at all.

We therefore make the following definition.

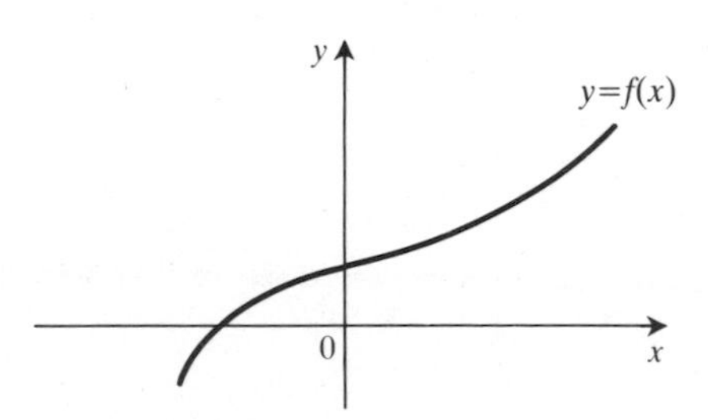

Fig. 4.9

4.18 Definition

The function $f(x)$ is *continuous* at the point $x = c$ if $\lim_{x\to c} f(x) = f(c)$. □

It is easy to prove on the basis of this definition that x^2, x^3, $1/(1+x^2)$ are continuous at every point, and that $\operatorname{sgn} x$ and $1/x$ are continuous everywhere except at $x = 0$. The function $\sin 1/x$ is clearly discontinuous at $x = 0$, and we shall shortly be able to prove it is continuous at all other points.

We define one-sided continuity by saying e.g. $f(x)$ is continuous *on the right* at $x = c$ if $\lim_{x\to c+} f(x) = f(c)$. For example, $f(x) = \sqrt{x}$ is continuous on the right at $x = 0$ since $\lim_{x\to 0+} \sqrt{x} = 0 = \sqrt{0}$. (See 4.10.) In fact, the function $\sqrt{x}$ is continuous at every $x > 0$ as well because of the inequality

$$|\sqrt{x} - \sqrt{y}| \leqslant \sqrt{|x - y|}$$

for all $x > 0$, $y > 0$. (See 2.39, question 3.)

We must now address ourselves to the question as to which functions are continuous. Also, having built up a stock of continuous functions, we must then ask what we can do with them.

The simplest way to manufacture continuous functions is by means of the operations of arithmetic, in which connection we have the following theorem.

4.19 Theorem

If $f(x)$, $g(x)$ are continuous at $x = c$, then so are
 (i) $f(x) + g(x)$,
 (ii) $f(x)g(x)$,
 (iii) $f(x)/g(x)$,
provided, in case (iii), $g(c) \neq 0$.

Proof is immediate from 4.5. The condition $g(c) \neq 0$ ensures that $g(x) \neq 0$ for all x belonging to some open interval (a, b) containing c. This is because there must be a $\delta > 0$ such that

$$|g(x) - g(c)| < |g(c)|$$

for all x satisfying $|x - c| < \delta$. (Play $\varepsilon = |g(c)|$.) Hence $g(x) \neq 0$ for all $x \in (c - \delta, c + \delta)$.

It follows that $f(x)/g(x)$ is defined for all $x \in (c - \delta, c + \delta)$, and it is clear that, for the purposes of evaluating $\lim_{x\to c} f(x)/g(x)$, it is enough to consider what happens on this interval. □

4.20 Corollaries

If $f(x)$, $g(x)$ are continuous at $x = c$, then so are
(i) $f(x) - g(x)$,
(ii) $Cf(x)$,
where C is any constant. □

4.21 Applications

(i) Any polynomial, e.g.,

$$p(x) = 5x^7 - 2x^6 + x^2 - 3$$

is continuous for every x.

In fact, $f(x) = \text{constant}$, $g(x) = x$ are clearly continuous from the definition (4.18), and $p(x)$ is obtained from $f(x)$, $g(x)$ by addition and multiplication.

(ii) Any rational function $p(x)/q(x)$ where $p(x)$, $q(x)$ are polynomials, e.g.

$$\frac{p(x)}{q(x)} = \frac{x^3 + 2x^2 + 3x + 4}{5x^3 + 6x^2 + 7x + 8},$$

is continuous everywhere except where the denominator $q(x)$ vanishes. □

Another way to combine continuous functions is by *composition*, or forming *composites*, defined as follows.

4.22 Definition

The *composition* or *composite* of two functions $f(x)$, $g(x)$ is the function $h(x) = f(g(x))$. □

For example (Fig. 4.10), if $f(x) = 1 + x$, $g(x) = |x|$, then

$$f(g(x)) = 1 + |x|,$$
$$g(f(x)) = |1 + x|.$$

Observe that $f(g(x))$, $g(f(x))$ are different functions in this case.

4.23 Theorem

If $g(x)$ is continuous at $x = c$, and $f(y)$ is continuous at $y = g(c)$, then $f(g(x))$ is continuous at $x = c$.

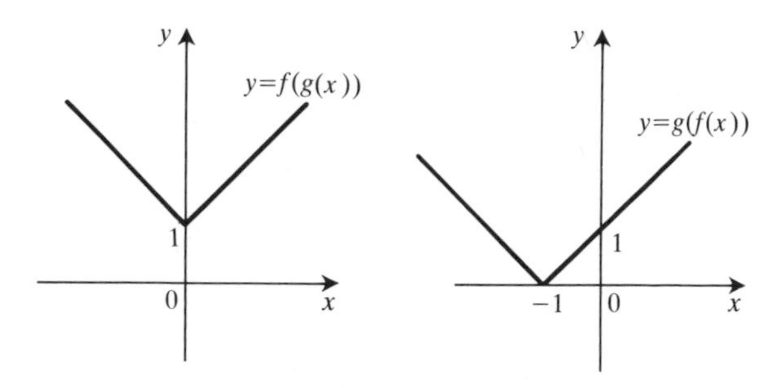

Fig. 4.10

Proof Let (x_n) be any sequence which converges to c. Then $g(x_n)$ converges to $g(c)$ by the continuity of $g(x)$ at $x=c$, and therefore $f(g(x_n))$ converges to $f(g(c))$ by the continuity of $f(y)$ at $y=g(c)$.

□

Hence, for example, the functions $1+|x|$, $|1+x|$ are continuous for all x. The function $|x|$ is of course continuous for all x since, e.g., $|x_n|\to|c|$ whenever $x_n\to c$. (See 2.39, question 2.)

The next method we shall consider of constructing new continuous functions from old ones is by taking *inverses*. Consider the function $f(x)=3x+4$. If we write

$$y=3x+4,$$

then we can solve for x in terms of y to obtain

$$x=\tfrac{1}{3}(y-4).$$

We shall call the function giving x in terms of y the *inverse* function of $f(x)$ and denote it by $f^{-1}(y)$, so that we have

$$f^{-1}(y)=\tfrac{1}{3}(y-4).$$

Observe that $f^{-1}(y)=x$ if and only if $f(x)=y$.

Constructing inverse functions is not always as easy as in the above example. This is because the equation $y=f(x)$ does not in general have a *unique* solution in x for *every* given y. Consider instead the example $f(x)=x^2$. If we try to solve $y=x^2$ in this case we find that, for $y>0$, there are two solutions, namely $x=\pm\sqrt{y}$, and, for $y<0$, there are no solutions for x. If we want to define an inverse function $f^{-1}(y)$ for $f(x)=x^2$, we shall have to restrict its domain of definition to $y\geqslant 0$ and, for $y>0$, decide which square root we shall take. The usual convention is to take $f^{-1}(y)$ equal to the positive square root of y, which we denote by $\sqrt{y}$. The negative square root is then of course denoted by $-\sqrt{y}$.

For a general function $f(x)$, we ensure *existence* of a solution x of the equation $y = f(x)$ for given y by means of the intermediate value theorem (see 4.24), and we ensure *uniqueness* of the solution by requiring $f(x)$ to be 'strictly monotonic' (see 4.25).

4.24 Intermediate value theorem

If $f(x)$ is continuous at every x in the closed interval $[a, b]$, and if γ satisfies $f(a) < \gamma < f(b)$, then there exists $c \in (a, b)$ such that $f(c) = \gamma$. □

N.B. If $f(x)$ is defined only for $x \in [a, b]$ we assume continuity on the right at a and on the left at b.

The intermediate value theorem is of course highly plausible—some might say downright obvious. Consider, however, the case $f(x) = x^2$, $[a, b] = [1, 2]$, $\gamma = 2$. We must have $c = \sqrt{2}$, which is of

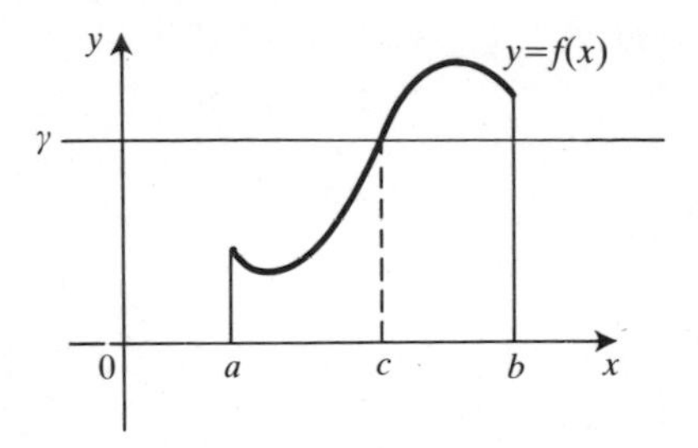

Fig. 4.11

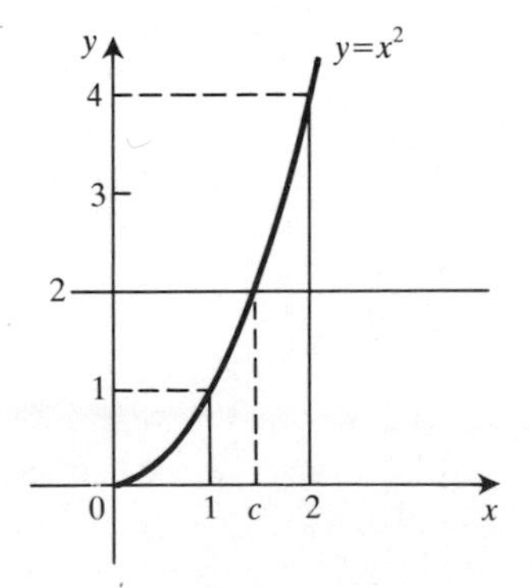

Fig. 4.12

course irrational. It follows that the intermediate value theorem is *false* in the system of rational numbers. It also follows that, to prove the theorem in the real number system, we should expect to have to appeal to the upper bound axiom, as indeed we shall.

Proof of 4.24 Let E be the set

$$E = \{x \in [a, b] : f(x) < \gamma\}.$$

E is non-empty since $a \in E$, and is bounded above since b is an upper bound. Therefore, by the upper bound axiom, E has a supremum. Let $c = \sup E$. We shall show $f(c) = \gamma$.

Choose sequences $x_n \in E$, $y_n \notin E$ such that

$$\lim_{n\to\infty} x_n = \lim_{n\to\infty} y_n = c.$$

(See 2.39, question 8.) Then, since $f(x)$ is continuous at $x = c$, we have

$$\lim_{n\to\infty} f(x_n) = \lim_{n\to\infty} f(y_n) = f(c).$$

But, since $x_n \in E$, $y_n \notin E$, we also have

$$f(x_n) < \gamma \leqslant f(y_n).$$

Therefore by the sandwich principle (2.16), we obtain $f(c) = \gamma$ as required. □

We should note that it is clear that $c \in [a, b]$ since $a \in E$ and b is an upper bound of E. It is clear also that we can assume $y_n \in [a, b]$ since, if $c < b$ there is no problem, and if $c = b$ then $c \notin E$, so we can take $y_n = b = c$ for all n.

4.25 Definition

We say $f(x)$ is *strictly increasing* over the interval $[a, b]$ if $f(x) < f(y)$ for every $x < y$ both $\in [a, b]$. We say $f(x)$ is *strictly decreasing* if $f(x) > f(y)$ whenever $x < y$. □

For example, $f(x) = \sin x$ is strictly increasing over $[-\frac{1}{2}\pi, \frac{1}{2}\pi]$, and strictly decreasing over $[\frac{1}{2}\pi, \frac{3}{2}\pi]$ (Fig. 4.13).

We say $f(x)$ is *strictly monotonic* over $[a, b]$ if *either* $f(x)$ strictly increases over $[a, b]$ *or* $f(x)$ strictly decreases over $[a, b]$.

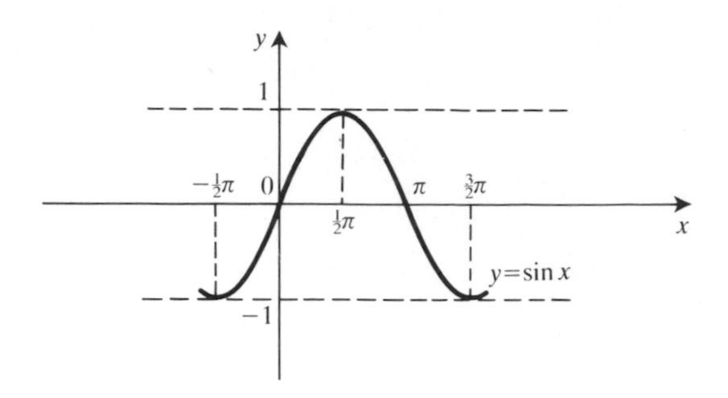

Fig. 4.13

4.26 Inverse function theorem

If $f(x)$ is continuous and strictly increasing over an interval $[a, b]$, then an inverse function $f^{-1}(y)$, such that $f^{-1}(y)=x$ whenever $f(x)=y$, can be defined for all $y \in [\alpha, \beta]$, where $\alpha=f(a)$, $\beta=f(b)$. Also, $f^{-1}(y)$ is continuous for every $y \in [\alpha, \beta]$.

Proof For every $y \in [\alpha, \beta]$, existence of x such that $f(x)=y$ follows from the intermediate value theorem (4.24), and uniqueness of this x follows from the fact that $f(x)$ is strictly increasing. Therefore the inverse function $f^{-1}(y)$ is certainly defined. It only remains to show that $f^{-1}(y)$ is continuous.

Suppose the sequence (y_n) converges to y. We must show that $f^{-1}(y_n)$ converges to $f^{-1}(y)$. Let $x_n=f^{-1}(y_n)$, $x=f^{-1}(y)$. We shall show firstly that (x_n) must converge, and then secondly that (x_n) must converge to x.

Suppose (x_n) were to diverge. Then, since (x_n) is bounded, it must have two subsequences $(x_{m_r})_{r\geq 1}$, $(x_{n_r})_{r\geq 1}$ which converge to

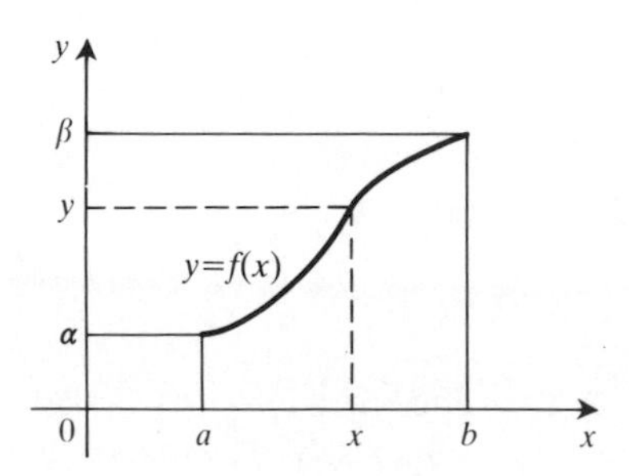

Fig. 4.14

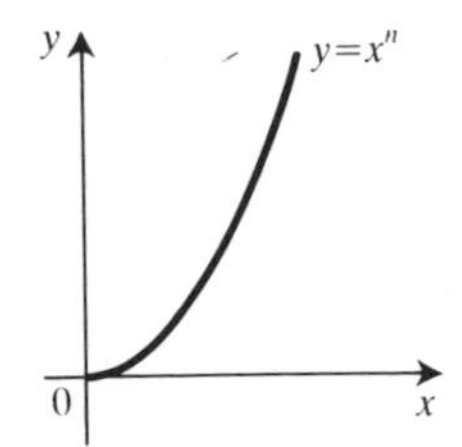

Fig. 4.15

distinct limits $l \neq l'$. (See 2.39, question 10.) Therefore, by continuity, $f(x_{m_r}) \to f(l)$, $f(x_{n_r}) \to f(l')$. However, $f(x_{m_r}) = y_{m_r}$, $f(x_{n_r}) = y_{n_r}$ are both subsequences of (y_n), which converges to y. Hence $f(l) = f(l') = y$, which contradicts $l \neq l'$.

So (x_n) must converge. Suppose the limit is l. Then by continuity, $f(x_n) \to f(l)$. But $f(x_n) = y_n \to y$. Therefore $y = f(l)$, and hence $l = x$. □

4.27 Application: *n*th roots

For each positive integer $n \geqslant 2$, the function x^n (Fig. 4.15) is continuous and strictly increasing over $x \geqslant 0$. Therefore, by 4.26, there is a well-defined inverse function, which we shall denote by $y^{1/n}$ and call the *positive nth root* of y (Fig. 4.16). The domain of definition of $y^{1/n}$ is $y \geqslant 0$ and it is continuous for all these y.

For n even (Fig. 4.17), there is of course a negative nth root as well. For n odd (Fig. 4.18), there is only one (real) nth root, but in this case there is an nth root of negative numbers also. □

The last method we shall describe for generating continuous functions involves the use of infinite series. Consider the following example.

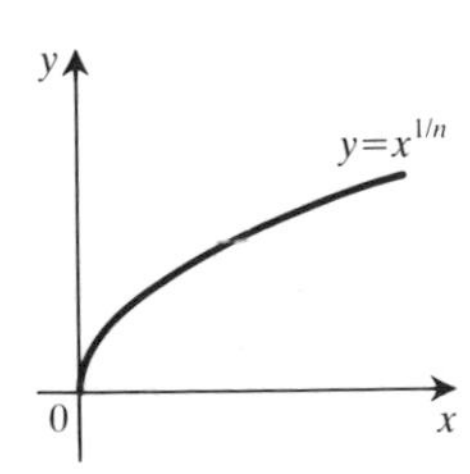

Fig. 4.16

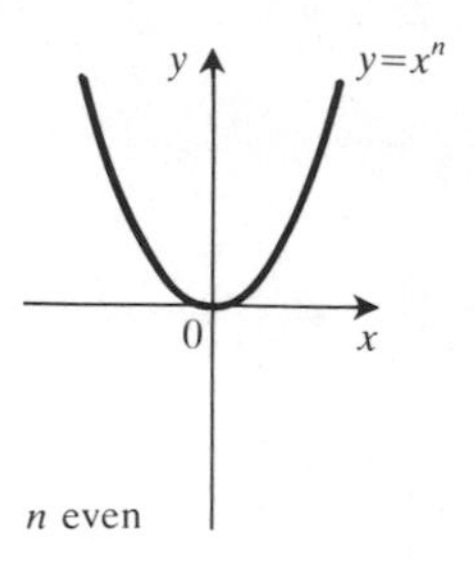

Fig. 4.17

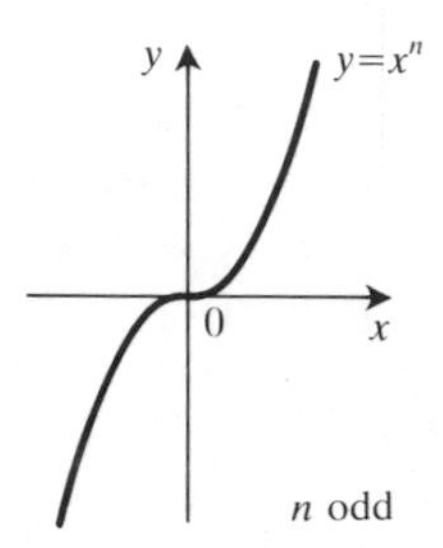

Fig. 4.18

4.28 Example

$$\sum_{0}^{\infty} x^n = \frac{1}{1-x} \qquad (|x|<1).$$

We can regard the series $\sum_0^\infty x^n$ as a series of *functions* whose nth term is the function x^n. The sum of the series is another function $1/(1-x)$. Observe that x^n is continuous for all $|x|<1$, and so is the sum function $1/(1-x)$. □

In general, however, we have no right to expect that the sum of an infinite series $\sum_1^\infty f_n(x)$ of continuous functions $f_n(x)$ is continuous. By induction on 4.19 (i), we can certainly say that the *partial* sums $\sum_1^N f_n(x)$ are continuous, but in the general situation we cannot say more than this. The following example gives an indication of the kind of thing that may happen (see Fig. 4.19).

4.29 Abel's example

$$\begin{aligned} \sum_{1}^{\infty} \frac{\sin nx}{n} &= \tfrac{1}{2}(\pi - x) \qquad (0<x<2\pi), \\ &= 0 \qquad (x=0,\ 2\pi). \end{aligned}$$ □

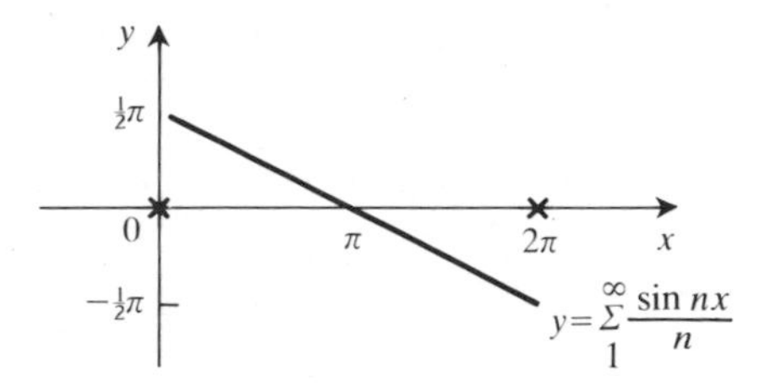

Fig. 4.19

We cannot give a proof of this result at this stage. (See Appendix, however.)

Observe that $(\sin nx)/n$ is continuous for all $x \in [0, 2\pi]$ but that the sum function has discontinuities at $x = 0, 2\pi$.

The historical importance of this example is that Abel published it in 1829 as a counter-example to an assertion by Cauchy that $\sum_1^\infty f_n(x)$ is always continuous if $f_n(x)$ is continuous for all n. In view of Abel's subsequent career, one cannot help wondering whether he should have been a little more tactful.

It turns out that we can say $\sum_1^\infty f_n(x)$ is continuous if we assume a little more about $f_n(x)$. In fact we have the following theorem.

4.30 Weierstrass' theorem

If $(f_n(x))_{n \geqslant 1}$ is a sequence of continuous functions on an interval $[a, b]$, and if there exists a convergent series $\sum_1^\infty M_n$ of constants M_n such that $|f_n(x)| \leqslant M_n$ for all $n \geqslant 1$ and for all $x \in [a, b]$, then the series $\sum_1^\infty f_n(x)$ converges for every $x \in [a, b]$, and its sum $s(x) = \sum_1^\infty f_n(x)$ is continuous on $[a, b]$. □

We shall call the condition that M_n with the above properties exists the *W-condition* (after Weierstrass). Observe that Abel's series $\sum_1^\infty (\sin nx)/n$ doesn't satisfy the W-condition over $[0, 2\pi]$.

Proof of 4.30 It is clear from the comparison test (3.13) that $\sum_1^\infty f_n(x)$ is absolutely convergent for each $x \in [a, b]$, so the sum function $s(x) = \sum_1^\infty f_n(x)$ is well defined over the interval $[a, b]$. We have to show it is continuous.

Suppose $c \in [a, b]$ and $\varepsilon > 0$ are given. Let N be chosen such that

$$\sum_{N+1}^{\infty} M_n < \tfrac{1}{3}\varepsilon$$

(see 3.38, question 16), and for each $n = 1, 2, \ldots, N$, let $\delta_n > 0$ be chosen such that

$$|f_n(x) - f_n(c)| < \frac{\varepsilon}{3N}$$

for all $|x - c| < \delta_n$. Then, if we set $\delta = \min_{1 \leq n \leq N} \delta_n$, we have $\delta > 0$ and, for any $|x - c| < \delta$,

$$\begin{aligned}
|s(x) - s(c)| &= \left| \sum_1^\infty f_n(x) - \sum_1^\infty f_n(c) \right| \\
&= \left| \sum_1^N (f_n(x) - f_n(c)) + \sum_{N+1}^\infty f_n(x) - \sum_{N+1}^\infty f_n(c) \right| \\
&\leq \sum_1^N |f_n(x) - f_n(c)| + \sum_{N+1}^\infty |f_n(x)| + \sum_{N+1}^\infty |f_n(c)| \\
&< \sum_1^N \frac{\varepsilon}{3N} + \tfrac{1}{3}\varepsilon + \tfrac{1}{3}\varepsilon \\
&= \varepsilon.
\end{aligned}$$

Hence $s(x)$ is continuous at $x = c$ as required. □

4.31 Application

If the power series $\sum_0^\infty a_n x^n$ has radius of convergence R (see 3.33), then its sum is continuous for all $|x| < R$.

$-r$ r

$-R$ 0 R

Proof We can show $\sum_0^\infty a_n x^n$ satisfies the W-condition on every closed interval $[-r, r]$ where $r < R$. In fact, for all $|x| \leq r$ we have

$$|a_n x^n| \leq |a_n| r^n,$$

and $\sum_0^\infty a_n r^n$ is absolutely convergent. (See proof of 3.33.) □

With the aid of 4.31 we can now substantially add to our stock of continuous functions. We shall use power series to give a rigorous definition of the exponential function e^x, which of course includes a definition of e as e^1. We shall define $\log_e x$ as the inverse function of e^x, and then use these two functions to define general powers and logarithms. We shall also give a rigorous definition of the trigonometric functions $\sin x$, $\cos x$ via power series, and obtain some of their basic properties, including their periodicity, and a definition of π in terms of solutions of the equation $\cos x = 0$.

4.32 Definition

We define the *exponential function* $e(x)$ to be

$$e(x)=\sum_{0}^{\infty}\frac{x^n}{n!}=1+x+\frac{x^2}{2!}+\cdots \qquad (0!=1). \qquad \square$$

The power series has radius of convergence $R=\infty$, so $e(x)$ is defined and continuous for all x by 4.31.

4.33 Theorem: Exponential theorem

$e(x+y)=e(x)e(y)$. $\square$

Proof $e(x)=\Sigma_0^\infty x^n/n!$, $e(y)=\Sigma_0^\infty y^n/n!$ are both absolutely convergent, so, by 3.36, we have

$$\begin{aligned} e(x)e(y) &= \left(1+x+\frac{x^2}{2!}+\cdots\right)\left(1+y+\frac{y^2}{2!}+\cdots\right)\\ &= 1+x+y+\frac{x^2}{2!}+xy+\frac{y^2}{2!}+\cdots\\ &= 1+(x+y)+\frac{(x+y)^2}{2!}+\cdots\\ &= e(x+y). \end{aligned}$$

4.34 Definition

$e=e(1)=\Sigma_0^\infty(1/n!)=1+1+(1/2!)+\cdots$. $\square$

It can be proved that $e=\lim_{n\to\infty}(1+1/n)^n$. (See 4.59, question 4.)

4.35 Theorem

For x rational, $e(x)=e^x$. $\square$

Proof $e(0)=1=e^0$ is immediate from the definition of $e(x)$. $e(n)=(e(1))^n=e^n$ for all positive integers n follows from 4.33. Also

$$\left(e\left(\frac{1}{n}\right)\right)^n=e\left(\frac{n}{n}\right)=e(1)=e,$$

and clearly $e(1/n)>0$, so $e(1/n)=e^{1/n}$ for all positive integers n. (See 4.27.)

If m, n are positive integers $\geqslant 2$, then

$$\mathrm{e}\left(\frac{m}{n}\right)=\left(\mathrm{e}\left(\frac{1}{n}\right)\right)^{m}=(\mathrm{e}^{1/n})^{m}=\mathrm{e}^{m/n}.$$

Finally, for any x we have

$$\mathrm{e}(x)\mathrm{e}(-x)=\mathrm{e}(x-x)=\mathrm{e}(0)=1.$$

Therefore

$$\mathrm{e}(-x)=\frac{1}{\mathrm{e}(x)}=\frac{1}{\mathrm{e}^{x}}=\mathrm{e}^{-x}$$

if x is rational. □

4.36 Definition

For x irrational we define $\mathrm{e}^x=\mathrm{e}(x)$. □

Theorem 4.33 now takes on the more familiar form

$$\mathrm{e}^{x+y}=\mathrm{e}^x\mathrm{e}^y.$$

4.37 Theorem

(i) e^x increases strictly for all x.
(ii) $\mathrm{e}^x\to\infty$ as $x\to\infty$.
(iii) $\mathrm{e}^x\to 0$ as $x\to-\infty$. □

Proof The fact that $\mathrm{e}^x\mathrm{e}^{-x}=\mathrm{e}^{x-x}=\mathrm{e}^0=1$ shows that $\mathrm{e}^x\neq 0$ for any x. Therefore $\mathrm{e}^x>0$ for all x since, if e^x were ever negative, then, by the intermediate value theorem (4.24), it would have to vanish somewhere.

It is clear from the definition 4.32 that $\mathrm{e}^x>1$ for all $x>0$. Therefore if $x<y$ we have

$$\mathrm{e}^y=\mathrm{e}^{y-x}\mathrm{e}^x>\mathrm{e}^x,$$

since $y-x>0$ and $\mathrm{e}^x>0$. This shows e^x is strictly increasing.

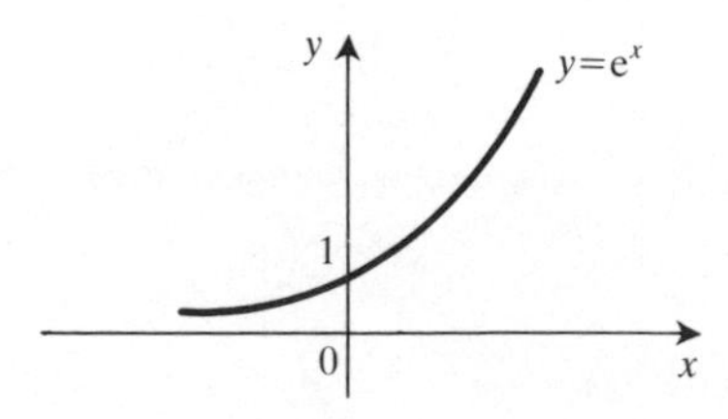

Fig. 4.20

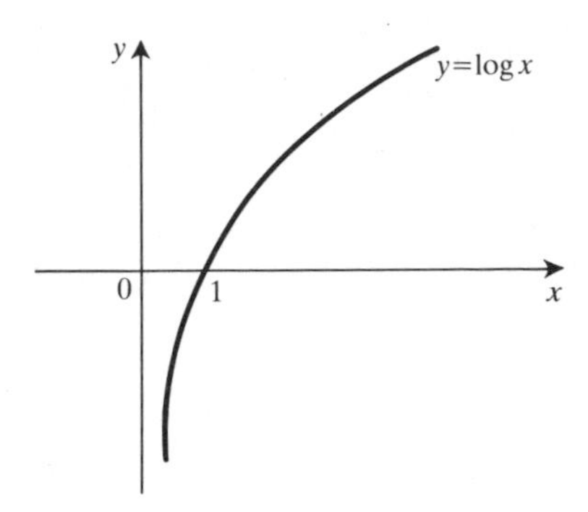

Fig. 4.21

To show $e^x \to \infty$ as $x \to \infty$, it is sufficient to observe that $e^x > x$ for all $x > 0$, another fact which is clear from 4.32.

It follows immediately that $e^x \to 0$ as $x \to -\infty$, since $e^x = 1/e^{-x}$ and $-x \to \infty$. □

4.38 Definition

For $x > 0$, we define the *logarithm* to the base e of x, denoted by $\log_e x$, or just $\log x$, to be the inverse function of e^x. □

The inverse function theorem (4.26) guarantees that $\log x$ is continuous for all $x > 0$. Clearly also $\log x$ is strictly increasing, tends to infinity as $x \to \infty$ and tends to minus infinity as $x \to 0_+$.

4.39 Theorem

$\log xy = \log x + \log y$. □

Proof Let $X = \log x$, $Y = \log y$, $Z = \log xy$. Then we have $e^X = x$, $e^Y = y$, $e^Z = xy$. Therefore

$$e^Z = e^X e^Y = e^{X+Y}$$

and hence $Z = X + Y$ as required. □

4.40 Theorem

For x rational and $a > 0$

$$\log a^x = x \log a.$$ □

Proof 4.40 is clearly true for $x = 0, 1$.

For any positive integer $n \geq 2$ we have

$$\log a^n = n \log a$$

by 4.39, also

$$n \log a^{1/n} = \log a,$$

and so

$$\log a^{1/n} = \frac{1}{n} \log a.$$

Therefore, if m, n are positive integers,

$$\log a^{m/n} = m \log a^{1/n} = \frac{m}{n} \log a.$$

Finally, if x is a negative rational, then

$$\log a^x + \log a^{-x} = \log (a^x a^{-x}) = \log 1 = 0,$$

and so

$$\log a^x = -\log a^{-x} = x \log a.$$ □

4.41 Definition

For x irrational and $a > 0$

$$a^x = e^{x \log a}.$$ □

Observe that this formula is a *theorem* for x rational (4.40), but a *definition* for x irrational.

Theorem 4.23 shows a^x is a continuous function of x (and a). Theorem 4.37 shows a^x is strictly increasing if $a > 1$ (Fig. 4.22),

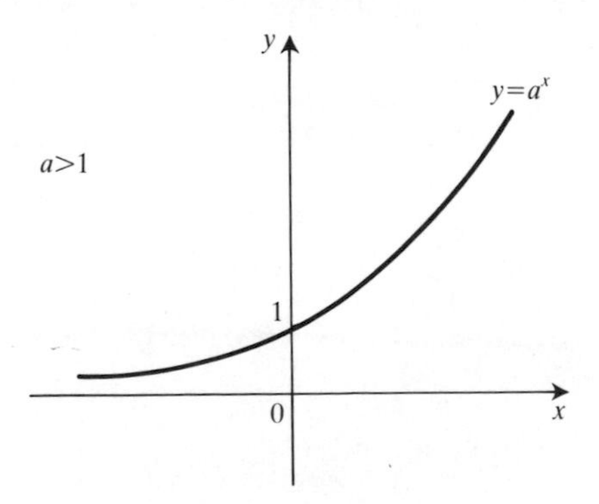

Fig. 4.22

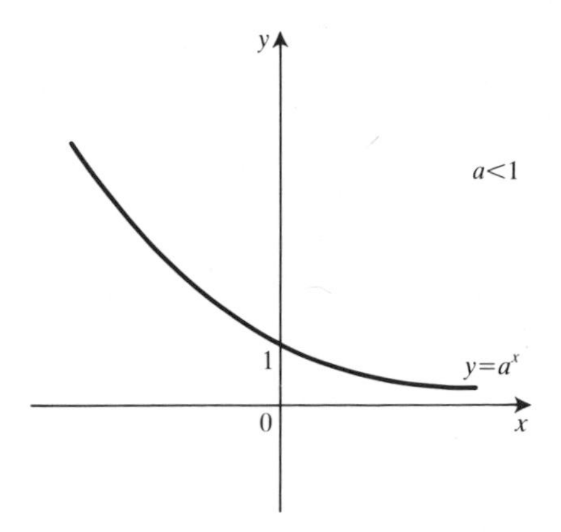

Fig. 4.23

strictly decreasing if $a < 1$ (Fig. 4.23), and takes on all positive values in both cases.

4.42 Exercise

Prove the laws of indices from the definition 4.41, i.e.,

(i) $a^{x+y} = a^x a^y$,
(i) $a^{xy} = (a^x)^y$,
(iii) $(ab)^x = a^x b^x$. □

4.43 Definition

For $a > 0$, $x > 0$ we define $\log_a x$ to be the inverse function of a^x. □

4.44 Exercise

Prove the laws of logarithms, i.e.,

(ii) $\log_a xy = \log_a x + \log_a y$,
(ii) $\log_a x^y = y \log_a x$,
(iii) $\log_a b \log_b c = \log_a c$. □

So much for powers and logarithms. Now for the trigonometric functions.

As promised we define the sine and cosine functions by means of their Maclaurin series as follows.

4.45 Definition

For all x we define

$$\sin x = x - \frac{x^3}{3!} + \frac{x^5}{5!} - \frac{x^7}{7!} + \cdots,$$

$$\cos x = 1 - \frac{x^2}{2!} + \frac{x^4}{4!} - \frac{x^6}{6!} + \cdots. \qquad \square$$

For the time being we must pretend that we know nothing about $\sin x$, $\cos x$ and build up their properties from the definitions 4.45 as if we had never seen these functions before. Of course we know what to expect from geometrical considerations, but an *arithmetical* approach demands that we sham ignorance, and appear genuinely surprised when the arithmetic bears out the geometry.

The first thing to observe is that both defining series are power series with infinite radius of convergence, and so $\sin x$, $\cos x$ are both defined and continuous for all x by 4.31.

4.46 Theorem: Addition formulae

$$\sin(x + y) = \sin x \cos y + \cos x \sin y$$

$$\cos(x + y) = \cos x \cos y - \sin x \sin y. \qquad \square$$

Proof The series for sine and cosine are both absolutely convergent for all values of the variable, and so it is legitimate to multiply them together by 3.36. Therefore we have

$$\begin{aligned}
&\sin x \cos y + \cos x \sin y \\
&= \left(x - \frac{x^3}{3!} + \cdots\right)\left(1 - \frac{y^2}{2!} + \cdots\right) + \left(1 - \frac{x^2}{2!} + \cdots\right)\left(y - \frac{y^3}{3!} + \cdots\right) \\
&= x + y - \frac{x^3}{3!} - \frac{xy^2}{2!} - \frac{x^2y}{2!} - \frac{y^3}{3!} + \cdots \\
&= (x + y) - \frac{(x + y)^3}{3!} + \cdots \\
&= \sin(x + y).
\end{aligned}$$

The second formula is proved similarly. $\square$

4.47 Theorem

$$\sin^2 x + \cos^2 x = 1. \qquad \square$$

Proof Put $y=-x$ in the cosine formula 4.46 and use the facts that $\cos 0=1$, $\cos(-x)=\cos x$, $\sin(-x)=-\sin x$, all easily provable from the definition 4.45. □

4.48 Corollary

$|\sin x|\leqslant 1$, $|\cos x|\leqslant 1$. □

4.49 Theorem

$\sin x>0$ for all $0<x<2$. □

Proof We have

$$\begin{aligned}\sin x &= \left(x-\frac{x^3}{3!}\right)+\left(\frac{x^5}{5!}-\frac{x^7}{7!}\right)+\cdots\\ &= \frac{x}{3!}(6-x^2)+\frac{x^5}{7!}(42-x^2)+\cdots\end{aligned}$$

which is clearly positive for all $0<x<2$.

4.50 Theorem

$\cos x$ strictly decreases for all $0\leqslant x\leqslant 2$. □

Proof If $0\leqslant x<y\leqslant 2$, then

$$\cos x-\cos y=2\sin\frac{x+y}{2}\sin\frac{y-x}{2}$$

(easily deducible from 4.46) which is positive by 4.49. □

4.51 Theorem

$\cos 2<0$. □

Proof We have

$$\begin{aligned}\cos 2 &= 1-\frac{2^2}{2!}+\frac{2^4}{4!}-\frac{2^6}{6!}+\frac{2^8}{8!}-\cdots\\ &= 1-2+\frac{2}{3}-\frac{2^6}{8!}(56-4)-\cdots\end{aligned}$$

which is clearly negative. □

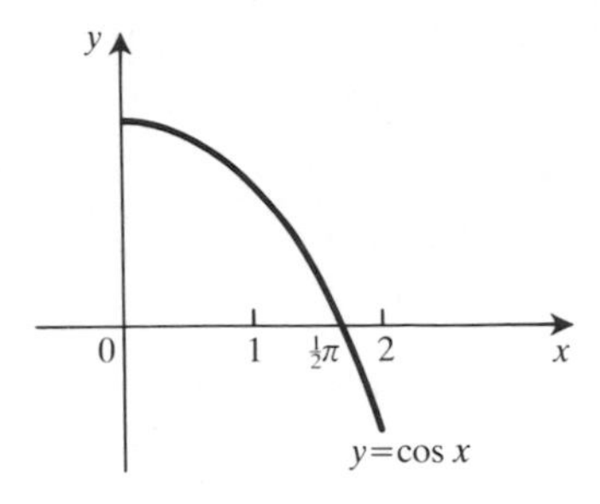

Fig. 4.24

4.52 Corollary

$\cos x$ vanishes precisely once in the interval $0<x<2$. □

Proof We have $\cos 0=1$ from the definition, so the result follows by the intermediate value theorem and the fact that $\cos x$ decreases strictly over $[0, 2]$. □

4.53 Definition of π

We define π to be the unique real number such that $0<\frac{1}{2}\pi<2$ and $\cos \frac{1}{2}\pi=0$ (see Fig. 4.24). □

4.54 Theorem

$\sin \frac{1}{2}\pi=1$. □

Proof Follows from 4.47 and 4.49. □

4.55 Corollaries

$$\sin\left(x+\frac{\pi}{2}\right)=\cos x, \qquad \cos\left(x+\frac{\pi}{2}\right)=-\sin x.$$ □

Proof Follows from 4.46, 4.53 and 4.54. □

4.55 enables the full behaviour of $\sin x$, $\cos x$ to be built up from the behaviour of $\cos x$ on $[0, \frac{1}{2}\pi]$. The graphs are as shown in Figs 4.25 and 4.26.

Observe that $\sin x$, $\cos x$ are both *periodic* with *period* 2π in the

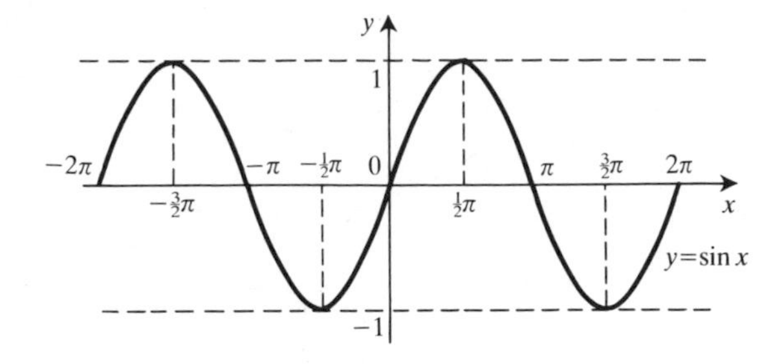

Fig. 4.25

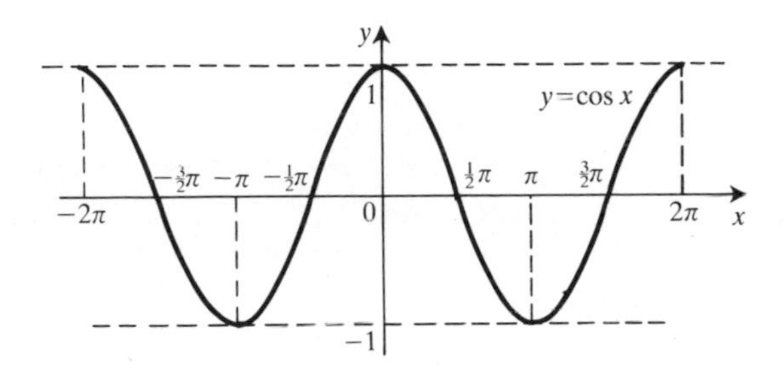

Fig. 4.26

sense that they repeat themselves every 2π, i.e.,

$$\sin(x+2\pi)=\sin x, \qquad \cos(x+2\pi)=\cos x.$$

Rigorous definitions of the other trigonometric functions can now be given by defining them in terms of sine and cosine. In fact, we have the following.

4.56 Definitions

(i) $\tan x = \dfrac{\sin x}{\cos x} \quad (x \neq (n+\frac{1}{2})\pi)$

(ii) $\cot x = \dfrac{\cos x}{\sin x} \quad (x \neq n\pi)$

(iii) $\sec x = \dfrac{1}{\cos x} \quad (x \neq (n+\frac{1}{2})\pi)$

(iv) $\operatorname{cosec} x = \dfrac{1}{\sin x} \quad (x \neq n\pi)$

We now have a sufficient supply of continuous functions, and know almost enough of their basic properties to be able to commence our treatment of the calculus. One property still needs to be considered, and that is the so-called 'minimax' property of the next theorem.

4.57 Definition

We say $f(x)$ is *bounded above* on an interval I if there exists M, called an *upper bound*, such that $f(x) \leq M$ for all $x \in I$. □

For example, $f(x) = \sin x$ is bounded above on $(0, \pi)$, but $f(x) = 1/x$ isn't.

We define bounded *below* and *lower* bound similarly, and say $f(x)$ is *bounded* if bounded above and below.

By the upper bound axiom, any $f(x)$ bounded above on an interval I must have a supremum, denoted by $\sup_{x \in I} f(x)$. This is a *maximum*, denoted by $\max_{x \in I} f(x)$, if and only if it is attained on I.

For example, $\sup_{0<x<\pi} \cos x = 1$, but there is no $x \in (0, \pi)$ such that $\cos x = 1$. On the other hand, $\max_{0<x<\pi} \sin x = 1$ exists since $\sin x = 1$ at $x = \frac{1}{2}\pi \in (0, \pi)$.

Similar remarks apply to $\inf_{x \in I} f(x)$ and $\min_{x \in I} f(x)$.

The minimax theorem says that under certain circumstances it is possible to assume that continuous functions always have maxima and minima.

4.58 Minimax theorem

If $f(x)$ is continuous on the bounded closed interval $[a, b]$, then $f(x)$ is bounded on $[a, b]$ and attains its supremum and infimum on $[a, b]$ (see Fig. 4.27). □

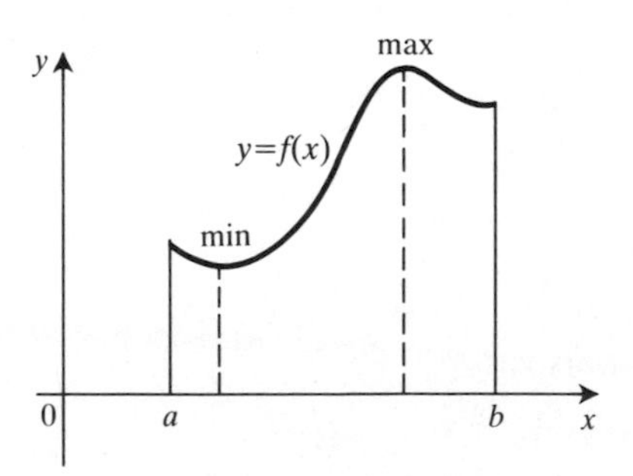

Fig. 4.27

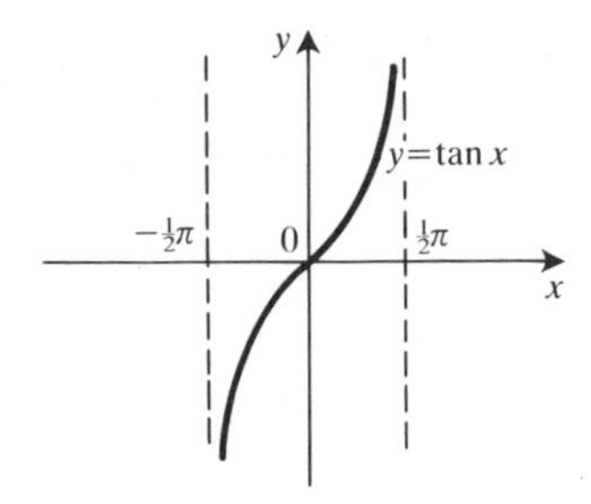

Fig. 4.28

Proof Assume (for contradiction) that $f(x)$ is unbounded above. Then (see 2.39, question 9) we can find a sequence (x_n) in $[a, b]$ such that $f(x_n)$ diverges to infinity. However, by the Bolzano–Weierstrass theorem (2.38), (x_n) must have a convergent subsequence $(x_{n_r})_{r\geq 1}$. Suppose $(x_{n_r})_{r\geq 1}$ converges to l. Then clearly $l \in [a, b]$ so, by continuity, $f(x_{n_r})$ must converge to $f(l)$. But $f(x_{n_r})$, being a subsequence of $f(x_n)$, must diverge to infinity (see 2.37). This is the required contradiction.

Let $M = \sup_{a\leq x\leq b} f(x)$. We want to show that there must exist $x \in [a, b]$ such that $f(x) = M$. We can certainly find a sequence (x_n) in $[a, b]$ such that $f(x_n) \to M$. (See 2.39, question 8.) By Bolzano–Weierstrass, there is a convergent subsequence $(x_{n_r})_{r\geq 1}$ with limit l say. Again, we clearly have $l \in [a, b]$ and therefore, by continuity, $f(x_{n_r}) \to f(l)$. But $f(x_{n_r})$is a subsequence of $f(x_n)$ which tends to M. Hence $f(l) = M$ as required.

The proofs that $f(x)$ is bounded below and attains its infimum on $[a, b]$ are similar. □

We should observe that the minimax theorem is no longer true if the interval $[a, b]$ is replaced by an unbounded or an unclosed interval. For example, $\tan x$ is continuous on the open interval $(-\frac{1}{2}\pi, \frac{1}{2}\pi)$, but is unbounded above and below (see Fig. 4.28).

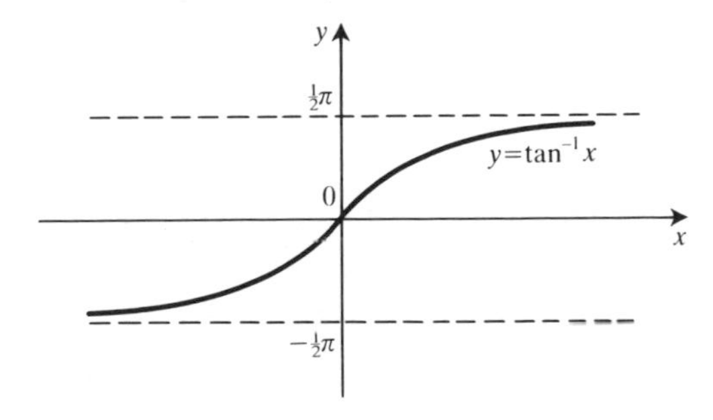

Fig. 4.29

Also $\tan^{-1} x$ is continuous for all x, taken as the inverse function of $\tan x$ on the interval $-\frac{1}{2}\pi < x < \frac{1}{2}\pi$, and, though it is bounded over all x, doesn't attain its supremum or infimum (see Fig. 4.29).

4.59 Miscellaneous exercises

1. Discuss the continuity of the following functions.

(i) $\operatorname{sgn} \sin x$ (ii) $\cos \operatorname{sgn} x$ (iii) $x \operatorname{sgn} x$

(iv) $f(x) = \operatorname{sgn} \sin \dfrac{1}{x}$ $(x \neq 0)$,
$\quad = 0$ $(x = 0)$.

(v) $f(x) = x \sin \dfrac{1}{x}$ $(x \neq 0)$,
$\quad = 0$ $(x = 0)$.

(vi) $f(x) = 1$ if x rational,
$\quad = 0$ if x irrational.

(vii) $f(x) = x$ if x rational,
$\quad = 1 - x$ if x irrational.

Draw graphs (where possible).

2. Evaluate the following limits.

(i) $\displaystyle\lim_{x\to 0} \frac{e^x - 1}{x}$ (ii) $\displaystyle\lim_{x\to 0} \frac{\sin x}{x}$ (iii) $\displaystyle\lim_{x\to 0} \frac{1 - \cos x}{x^2}$

Hint: Use 4.31.

3. Prove that

$$1 + x \leqslant e^x \leqslant \frac{1}{1 - x}$$

for all $|x| < 1$. *Hint:* Observe that $\sum_0^\infty x^n/n! \leqslant \sum_0^\infty x^n$ $(0 \leqslant x < 1)$.

4. Prove that

$$\frac{x}{1 + x} \leqslant \log(1 + x) \leqslant x$$

for all $|x| < 1$. *Hint:* Take logarithms in the inequalities of question 3.

Deduce that

$$\lim_{x\to 0} \frac{\log(1 + x)}{x} = 1.$$

Hence prove

$$\lim_{n\to\infty}\left(1+\frac{1}{n}\right)^n = \mathrm{e}.$$

Hint: Use 4.41, 4.3 and the continuity of e^x.

5. Prove that, for any positive integer n,

$$\lim_{x\to\infty}\frac{\mathrm{e}^x}{x^n}=\infty.$$

Hint: Compare $\sum_0^\infty x^n/n!$ with its $(n+1)$th term.

Deduce that

(i) $\lim_{x\to\infty} x^4\mathrm{e}^{-x}=0$,

(ii) $\lim_{n\to\infty} n^2\mathrm{e}^{-\sqrt{n}}=0$ (see remark after 4.16)

(iii) $\sum_1^\infty \mathrm{e}^{-\sqrt{n}}$ converges.

Observe: Exponentials dominate powers.

6. Prove that

(i) $\lim_{x\to\infty}\frac{\log x}{x}=0$, (ii) $\lim_{x\to 0} x\log x=0$.

Hint: Put $x=\mathrm{e}^t$ and let $t\to\pm\infty$. *Observe:* Powers dominate logarithms.

Discuss the convergence of $\sum_1^\infty(-1)^{n-1}(\log n)/n$.

7. Prove that

$$\left|1-\frac{\sin x}{x}\right|<\frac{x^2}{1-x^2}$$

for all $0<|x|<1$.

8. Prove that if $f(x)$ is continuous on $[0, 1]$ and $f(0)=1$, $f(1)=0$, then $f(x)=x$ for some $x\in(0, 1)$. *Hint:* Apply the intermediate value theorem to $g(x)=f(x)-x$.

9. Prove that if $f(x)$ is continuous on $[0, 1]$ and $f(x)\neq 0$ for all $x\in[0, 1]$, then $1/f(x)$ is bounded on $[0, 1]$. *Hint:* Use the minimax theorem to show $\inf_{0\leq x\leq 1}|f(x)|$ is positive.

10. Prove that, if $f(x)$ is continuous on $[a, b]$, and if $\varepsilon>0$ is given, then there exists a smallest $x\in[a, b]$ such that $|f(x)-f(a)|=\varepsilon$, unless $|f(x)-f(a)|<\varepsilon$ for all $x\in[a, b]$. *Hint:* Consider $\inf E$, where

$$E=\{x\in[a, b]:|f(x)-f(a)|=\varepsilon\}.$$

The intermediate value theorem shows E is not empty.

5
Differential calculus

We now wish to give a rigorous treatment of differential calculus. We shall give a precise definition of what it means to differentiate a function, and build a theory of differentiation on this definition. The most important theorem of the chapter is the so-called mean value theorem, which we shall establish rigorously and then put to a variety of applications.

Differentiating a function means finding its instantaneous rate of change. Graphically, it means finding the slope of the tangent to the curve $y = f(x)$ at a given point P (see Fig. 5.1).

To obtain a numerical value for the rate of change of $f(x)$ at the instant x, we consider the average rate of change over an interval $[x, x+h]$, i.e.,

$$\frac{f(x+h)-f(x)}{h},$$

and see what happens when $h \to 0$. This is equivalent to drawing a chord PQ on the graph and considering the limit of the slope of PQ as Q→P (see Fig. 5.2).

We are led to define the *derivative* $f'(x)$ of the function $f(x)$ as the limit

$$f'(x) = \lim_{h\to 0} \frac{f(x+h)-f(x)}{h}.$$

Of course we cannot assume that the limit will always exist, so we say $f(x)$ is *differentiable* whenever the limit does exist. We shall exclude the case where

$$\lim_{h\to 0} \frac{f(x+h)-f(x)}{h} = \pm\infty,$$

so that, if $f(x)$ is differentiable over an interval I, then $f'(x)$ is another function defined on I.

The notation $f'(x)$ will be used exclusively in preference to the Leibnitz notation dy/dx, though it will sometimes be useful to rephrase things the Leibnitz way in order to see what is going on in

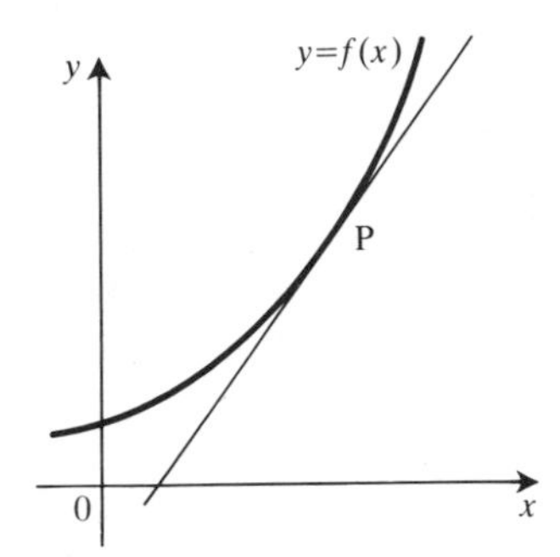

Fig. 5.1

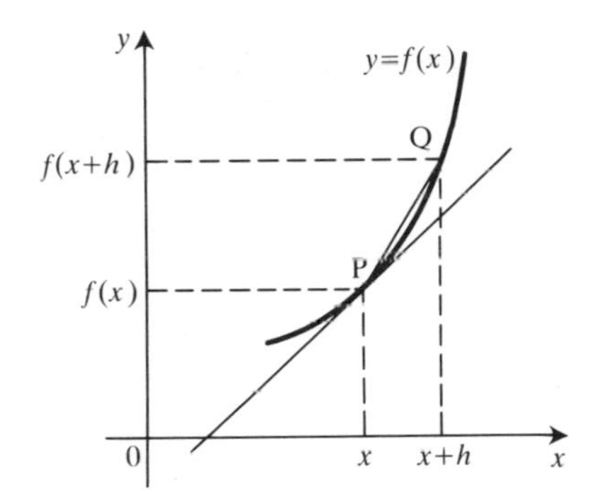

Fig. 5.2

a situation before attempting a rigorous justification. The expression dy/dx of course means the *limiting value* of the ratio between the change dy in y and the change dx in x as both dy and dx tend to zero.

The formal definition is as follows.

5.1 Definition

$f(x)$ is *differentiable* at x if

$$\lim_{h \to 0} \frac{f(x+h) - f(x)}{h}$$

exists (finite), and the value of the limit wherever it exists is called the *derivative* of $f(x)$ at x, and is denoted by $f'(x)$. □

5.2 Examples

(i) $f(x) \equiv C$ (constant).

In this case we have

$$\frac{f(x+h)-f(x)}{h}=\frac{C-C}{h}=0$$

for all x and for all $h \neq 0$, so $f(x)$ is differentiable at every x with derivative $f'(x) \equiv 0$.

(ii) $f(x)=x$.

Here we have

$$\frac{f(x+h)-f(x)}{h}=\frac{(x+h)-x}{h}=\frac{h}{h}=1$$

for all x and for all $h \neq 0$, so $f(x)$ is differentiable everywhere with derivative $f'(x) \equiv 1$.

(iii) $f(x)=1/x \quad (x \neq 0),$

$\qquad\quad = 0 \quad (x=0).$

If $x \neq 0$, then

$$\frac{f(x+h)-f(x)}{h}=-\frac{1}{x(x+h)}$$

for all h satisfying $0<|h|<|x|$. Therefore $f(x)$ is differentiable with $f'(x)=-1/x^2$ for all $x \neq 0$.

If $x=0$, then

$$\frac{f(x+h)-f(x)}{h}=\frac{f(h)-f(0)}{h}=\frac{1}{h^2}$$

which diverges to infinity as $h \to 0$, so $f(x)$ is not differentiable at $x=0$. □

The following theorem describes the fundamental relationship between differentiable functions and continuous functions.

5.3 Theorem

If a function is differentiable at any point, then it must be continuous there. □

Proof Suppose $f(x)$ is differentiable at x. Then

$$f(x+h)-f(x)=h\frac{f(x+h)-f(x)}{h}$$
$$\rightarrow 0 \, . \, f'(x)$$
$$=0$$

as $h \rightarrow 0$. Hence $f(x)$ is continuous at x. □

Observe that $1/x$ is discontinuous at $x=0$ whatever (finite) value we give it at $x=0$, so cannot be differentiable there, by 5.3, confirming our findings of 5.2 (iii).

Observe also that the converse of 5.3 is false. The standard counter-example is the following.

5.4 Counter-example

$f(x)=|x|$.

$f(x)$ is continuous at $x=0$, but is not differentiable there since

$$\frac{f(h)-f(0)}{h}=\frac{|h|}{h}=\operatorname{sgn} h \qquad (h \neq 0)$$

which has no limit as $h \rightarrow 0$. □

The above example leads us to introduce the notion of one-sided differentiability, which we define as follows.

5.5 Definition

$f(x)$ is differentiable *on the right* at x if

$$\lim_{h \rightarrow 0_+} \frac{f(x+h)-f(x)}{h}$$

exists (finite). The limit is called the *right-hand* derivative and is denoted by $f'_+(x)$. □

Left-hand differentiability and derivative are defined similarly.

For example, $f(x)=|x|$ has $f'_+(0)=1$, $f'_-(0)=-1$.

Our next task is to decide which functions are differentiable, and to find their derivatives. We shall adopt a similar procedure to that used in Chapter 4 of constructing new functions from old ones by various methods. Here we not only have to establish differentiability of the new functions, we also have to determine their derivatives in terms of the derivatives of the old functions.

5.6 Theorem: Arithmetic of differentiable functions

If $f(x)$, $g(x)$ are differentiable at x, then so are

(i) $s(x)=f(x)+g(x)$,
(ii) $p(x)=f(x)g(x)$,
(iii) $q(x)=f(x)/g(x)$,

provided, in case (iii), $g(x)\neq 0$, and their derivatives are

(i) $s'(x)=f'(x)+g'(x)$,
(ii) $p'(x)=f'(x)g(x)+f(x)g'(x)$,
(iii) $q'(x)=(f'(x)g(x)-f(x)g'(x))/(g(x))^2$.

Proofs (i) We have

$$\frac{s(x+h)-s(x)}{h}=\frac{f(x+h)-f(x)}{h}+\frac{g(x+h)-g(x)}{h}$$
$$\to f'(x)+g'(x)$$

as $h\to 0$. Hence $s(x)$ is differentiable at x with derivative

$$s'(x)=f'(x)+g'(x).$$

(ii) We have

$$\frac{p(x+h)-p(x)}{h}=\frac{f(x+h)g(x+h)-f(x)g(x)}{h}$$
$$=\frac{f(x+h)-f(x)}{h}g(x+h)+f(x)\frac{g(x+h)-g(x)}{h}$$
$$\to f'(x)g(x)+f(x)g'(x)$$

as $h\to 0$ (by 5.3).

(iii) If $g(x)\neq 0$, then also $g(x+h)\neq 0$ for all $|h|<\delta$ for some $\delta>0$ by continuity (5.3). Therefore, if h satisfies $0<|h|<\delta$ for this δ, then we have

$$\frac{q(x+h)-q(x)}{h}=\frac{f(x+h)g(x)-f(x)g(x+h)}{hg(x)g(x+h)}$$
$$=\frac{1}{g(x)g(x+h)}$$
$$\times\left\{\frac{f(x+h)-f(x)}{h}g(x)-f(x)\frac{g(x+h)-g(x)}{h}\right\}$$
$$\to\frac{f'(x)g(x)-f(x)g'(x)}{(g(x))^2}$$

as $h\to 0$, by 5.3. □

5.7 Corollaries

(i) If $f(x)$ is differentiable at x, then so is $Cf(x)$ for any constant C, and the derivative is $Cf'(x)$.

(ii) If $f(x)$, $g(x)$ are differentiable at x, then so is $f(x)-g(x)$, with derivative $f'(x)-g'(x)$.

Proofs (i) follows from 5.2 (i) and 5.6 (ii).
(ii) follows from (i) and 5.6 (i). □

5.8 Exercises

Prove that for any positive integer n,
(i) x^n is differentiable with derivative nx^{n-1},
(ii) $1/x^n$ is differentiable for all $x \neq 0$ with derivative $-n/x^{n+1}$.
Hint. Use 5.2, 5.6 and induction.

5.9 Applications

(i) Any polynomial $p(x)$ is differentiable for all x.

(ii) Any rational function $r(x)=p(x)/q(x)$, where $p(x)$, $q(x)$ are polynomials, is differentiable except where $q(x)=0$.

5.10 Theorem: Composition of differentiable functions

If $g(x)$ is differentiable at x, and if $f(y)$ is differentiable at $y=g(x)$, then $c(x)=f(g(x))$ is differentiable at x with derivative $c'(x)=f'(g(x))g'(x)$. □

Observe that Leibniz notation is suggestive here. If we write $z=f(y)$, we have $z=f(g(x))=c(x)$, and so

$$c'(x)=\frac{dz}{dx}=\frac{dz}{dy}\frac{dy}{dx}=f'(y)g'(x)=f'(g(x))g'(x)$$

on the assumption that dy can be cancelled. Two questions arise: firstly, can dy be cancelled in the limit? and secondly, how do we proceed if $dy=0$?

The formal proof is as follows.

Proof of 5.10 We distinguish two cases.

Case 1
$g'(x) \neq 0$

In this case there must exist $\delta > 0$ such that

$$\frac{g(x+h)-g(x)}{h} \neq 0,$$

and therefore $g(x+h) \neq g(x)$, for all $0 < |h| < \delta$. Hence for h in this range we have

$$\begin{aligned}\frac{c(x+h)-c(x)}{h} &= \frac{f(g(x+h))-f(g(x))}{h} \\ &= \frac{f(g(x+h))-f(g(x))}{g(x+h)-g(x)} \frac{g(x+h)-g(x)}{h} \\ &= \frac{f(y+k)-f(y)}{k} \frac{g(x+h)-g(x)}{h},\end{aligned}$$

where $k = g(x+h) - g(x) \neq 0$,

$$\begin{aligned}&\to f'(y)g'(x) \\ &= f'(g(x))g'(x)\end{aligned}$$

as $h \to 0$, since $k \to 0$ as $h \to 0$ by the continuity of $g(x)$.

Case 2
$g'(x) = 0$

The argument of case 1 shows that

$$\frac{c(x+h)-c(x)}{h} \to f'(g(x))g'(x)$$

as $h \to 0$ via values such that $g(x+h) \neq g(x)$. If $g(x+h) = g(x)$ for any $h \neq 0$, then we have

$$\frac{c(x+h)-c(x)}{h} = \frac{f(g(x+h))-f(g(x))}{h} = 0,$$

and hence

$$\frac{c(x+h)-c(x)}{h} \to 0 = f'(g(x))g'(x)$$

as $h \to 0$ via these values also. □

5.11 Examples

(i) $f(x) = \sin \frac{1}{x} \qquad (x \neq 0),$

$\qquad = 0 \qquad (x = 0).$

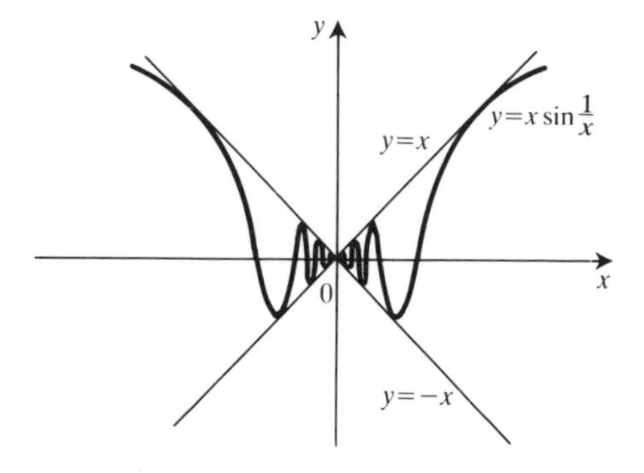

Fig. 5.3

$f(x)$ is differentiable for all x except $x=0$, where $f(x)$ is not differentiable since $f(x)$ is not continuous at $x=0$.

Of course we are assuming $\sin x$ is differentiable for all x. This will be justified shortly. (See 5.18.)

(ii) $f(x)=x\sin\dfrac{1}{x}$ $\quad(x\neq 0)$,

$\quad=0$ $\quad(x=0)$ (see Fig. 5.3).

$f(x)$ is continuous everywhere and differentiable everywhere except $x=0$. In fact

$$\frac{f(h)-f(0)}{h}=\sin\frac{1}{h}$$

which has no limit as $h\to 0$.

(iii) $f(x)=x^2\sin\dfrac{1}{x}$ $\quad(x\neq 0)$,

$\quad=0$ $\quad(x=0)$ (see Fig. 5.4).

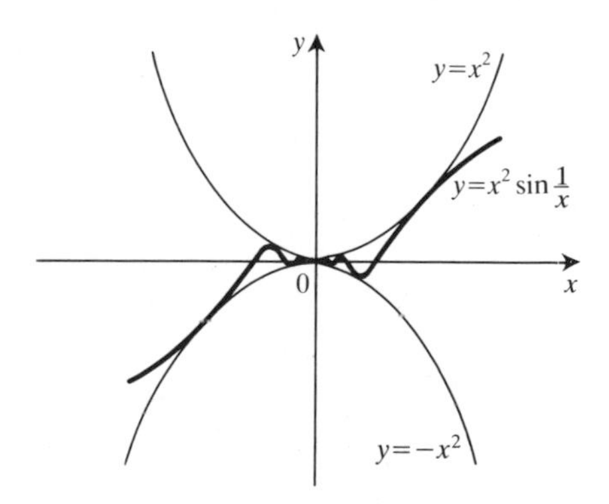

Fig. 5.4

$f(x)$ is differentiable everywhere including $x = 0$. In fact,

$$\frac{f(h) - f(0)}{h} = h \sin \frac{1}{h} \to 0$$

as $h \to 0$, so $f'(0)$ exists and $=0$.

5.12 Exercise

Find the derivative $f'(x)$ of the function

$$\begin{aligned} f(x) &= x^2 \sin \frac{1}{x} \qquad (x \neq 0) \\ &= 0 \qquad\qquad\quad (x = 0) \end{aligned}$$

at $x \neq 0$. (Assume the derivative of $\sin x$ is $\cos x$.) Show $f'(x)$ is discontinuous at $x = 0$. Explain why this doesn't contradict 5.3. □

5.13 Inverse function theorem

If $y = f(x)$ is differentiable and strictly increasing over an interval $[a, b]$, and if $f'(x) \neq 0$ for all $x \in [a, b]$, then the inverse function $x = \phi(y)$ is differentiable over $[f(a), f(b)]$ with derivative $\phi'(y) = 1/f'(x) = 1/f'(\phi(y))$. □

Again we are in a situation where it is illuminating to use Leibnitz notation. Here we have

$$\phi'(y) = \frac{dx}{dy} = 1 \bigg/ \frac{dy}{dx} = 1/f'(x)$$

and the question is whether we can invert dy/dx in the limit. The justification is as follows.

Proof of 5.13 For any $k \neq 0$ we have

$$\frac{\phi(y + k) - \phi(y)}{k} = \frac{h}{f(x + h) - f(x)}$$

where $x = \phi(y)$, $x + h = \phi(y + k)$. Now $k \to 0$ implies

$$h = \phi(y + k) - \phi(y) \to 0$$

by the continuity of $\phi(y)$. (See 4.26.) Therefore

$$\frac{\phi(y + k) - \phi(y)}{k} \to \frac{1}{f'(x)}$$

as $k \to 0$, which is the required result. □

5.14 Application: Differentiability of $x^{1/n}$ $(n \geqslant 2)$

Theorem 5.13 shows that $f(x) = x^{1/n}$ is differentiable for all $x > 0$ with derivative

$$f'(x) = \frac{1}{ny^{n-1}},$$

where $y = x^n$,

$$= \frac{1}{nx^{(n-1)/n}}$$

$$= \frac{1}{n} x^{(1/n)-1}.$$

If n is odd, $x^{1/n}$ is differentiable for all $x < 0$ also. At $x = 0$, $x^{1/n}$ has an infinite derivative, one-sided if n is even, two-sided if n is odd, which corresponds to the fact that the tangent to the graph is vertical at the origin. This indicates what happens when the condition $f'(x) \neq 0$ is left out of Theorem 5.13. □

The final method we shall consider for constructing differentiable functions is by means of infinite series. There is a Weierstrassian theorem (cf. 4.30) which enables differentiation of an infinite series of functions to be justified, but its proof relies on the mean value theorem, so it will have to wait until that theorem is covered.

We shall therefore proceed forthwith to the mean value theorem, and among its applications the first will be to prove Weierstrass' theorem for differentiable functions. We shall attack the mean value theorem via a special case of it known as Rolle's theorem, which is as follows.

5.15 Rolle's theorem

If $f(x)$ is continuous on $[a, b]$ and differentiable on (a, b), and if $f(a) = f(b)$, then there exists $c \in (a, b)$ such that $f'(c) = 0$ (see Fig. 5.5).

Proof If $f(x)$ is constant on $[a, b]$, then $f'(x) = 0$ for all $x \in [a, b]$. Otherwise $f(x)$ attains a maximum or minimum at an interior point $c \in (a, b)$ (see 4.58).

If $f(c) \geqslant f(x)$ for all $x \in [a, b]$, then

$$\frac{f(c+h) - f(c)}{h} \leqslant 0$$

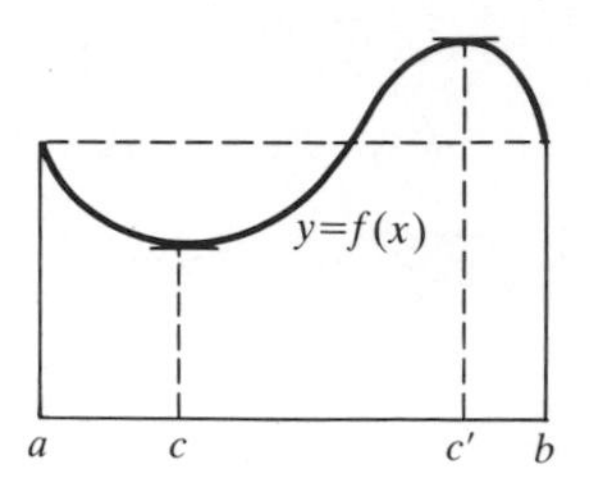

Fig. 5.5

if $h > 0$, and

$$\frac{f(c+h) - f(c)}{h} \geqslant 0$$

if $h < 0$, so letting $h \to 0$ we obtain $f'(c) \leqslant 0$, $f'(c) \geqslant 0$ simultaneously, and hence $f'(c) = 0$.

If $f(c) \leqslant f(x)$ for all $x \in [a, b]$, then $f'(c) = 0$ similarly. □

5.16 Mean value theorem

If $f(x)$ is continuous on $[a, b]$ and differentiable on (a, b), then

$$f'(c) = \frac{f(b) - f(a)}{b - a}$$

for some $c \in (a, b)$ (see Fig. 5.6) □

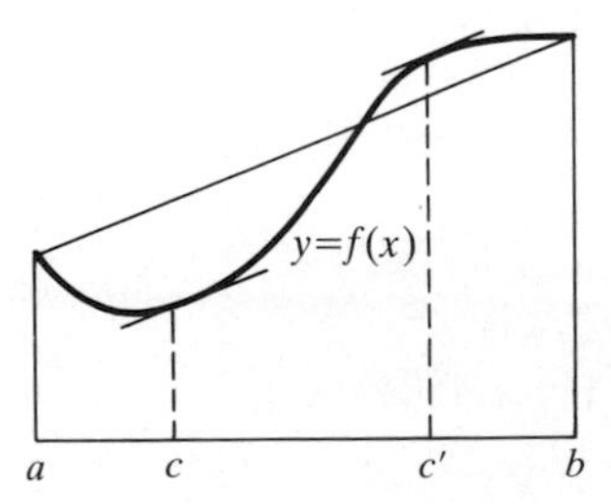

Fig. 5.6

Observe that the expression

$$\frac{f(b)-f(a)}{b-a}$$

represents the 'mean value' of $f'(x)$ over the interval $[a, b]$, and the mean value theorem is saying that $f'(x)$ must attain this mean value somewhere in the open interval (a, b).

Proof Let

$$g(x)=f(x)-Cx$$

where the constant C is chosen in such a way that $g(x)$ satisfies the conditions of Rolle's theorem. In fact, we require

$$f(a)-Ca=f(b)-Cb,$$

which gives

$$C=\frac{f(b)-f(a)}{b-a}.$$

Then, by Rolle's theorem, there must exist $c \in (a, b)$ such that $g'(c)=0$. Hence $f'(c)=C$ as required. □

An equivalent way of stating the mean value theorem is to say that if $f(x)$ is continuous over the interval $[x, x+h]$ and differentiable over the interval $(x, x+h)$, then

$$\frac{f(x+h)-f(x)}{h}=f'(x+\theta h)$$

for some θ satisfying $0<\theta<1$. Observe that this statement is equally valid for h positive or negative.

As promised, our first application of the mean value theorem is to the differentiation of infinite series.

5.17 Weierstrass' theorem

If $\sum_1^\infty f_n(x)$ is a convergent series of differentiable functions on the interval $[a, b]$, and if there exists a convergent series of positive constants $\sum_1^\infty M_n'$ such that

$$|f_n'(x)| \leqslant M_n'$$

for all $x \in [a, b]$ and for all $n \geqslant 1$ (*W'-condition*), then $s(x)=$

$\sum_1^\infty f_n(x)$ is differentiable on $[a, b]$ and its derivative is $s'(x) = \sum_1^\infty f_n'(x)$. □

Observe that Abel's series $\sum_1^\infty (\sin nx)/n$ (see 4.29) formally differentiates to $\sum_1^\infty \cos nx$, which doesn't satisfy the W′-condition. In fact, $\sum_1^\infty \cos nx$ diverges for every x.

Proof of 5.17 The W′-condition clearly implies immediately that $\sum_1^\infty f_n'(x)$ converges for each $x \in [a, b]$. Suppose $x \in [a, b]$ and $\varepsilon > 0$ are given. Then we can choose N such that

$$\sum_{N+1}^{\infty} M_n' < \tfrac{1}{3}\varepsilon,$$

and for each $n = 1, 2, \ldots, N$ we can choose $\delta_n > 0$ such that

$$\left|\frac{f_n(x+h) - f_n(x)}{h} - f_n'(x)\right| < \frac{\varepsilon}{3N}$$

for all $0 < |h| < \delta_n$. Let $\delta = \min_{1 \leqslant n \leqslant N} \delta_n$. Then $\delta > 0$ and, for any $0 < |h| < \delta$, we have

$$\begin{aligned}
&\left|\frac{s(x+h) - s(x)}{h} - \sum_{1}^{\infty} f_n'(x)\right| \\
&\quad = \left|\sum_{1}^{\infty} \left(\frac{f_n(x+h) - f_n(x)}{h} - f_n'(x)\right)\right| \\
&\quad \leqslant \sum_{1}^{N} \left|\frac{f_n(x+h) - f_n(x)}{h} - f_n'(x)\right| + \sum_{N+1}^{\infty} \left|\frac{f_n(x+h) - f_n(x)}{h} - f_n'(x)\right| \\
&\quad < \sum_{1}^{N} \frac{\varepsilon}{3N} + \sum_{N+1}^{\infty} |f_n'(x + \theta_n h) - f_n'(x)|,
\end{aligned}$$

for some $0 < \theta_n < 1$ by the mean value theorem,

$$\begin{aligned}
&\quad \leqslant \tfrac{1}{3}\varepsilon + \sum_{N+1}^{\infty} (|f_n'(x + \theta_n h)| + |f_n'(x)|) \\
&\quad < \tfrac{1}{3}\varepsilon + 2\sum_{N+1}^{\infty} M_n' \\
&\quad < \tfrac{1}{3}\varepsilon + \tfrac{2}{3}\varepsilon \\
&\quad = \varepsilon.
\end{aligned}$$

5.18 Application to power series

If the power series $\sum_0^\infty a_n x^n$ has radius of convergence R, then $f(x) = \sum_0^\infty a_n x^n$ is differentiable over $|x| < R$ and its derivative there is

$$f'(x) = \sum_{1}^{\infty} na_n x^{n-1}.$$

Proof The differentiated series $\sum_1^\infty na_n x^{n-1}$ also has radius of convergence R (see 3.38, question 17). Therefore, for any r satisfying $0 < r < R$, the series $\sum_1^\infty na_n r^{n-1}$ is absolutely convergent, and so $\sum_1^\infty a_n x^n$ satisfies the W′-condition over $[-r, r]$. (Take $M'_n = n\,|a_n|\,r^{n-1}$.) The result follows. □

We can now differentiate powers and logarithms. In fact, by definition,

$$e^x = 1 + x + \frac{x^2}{2!} + \cdots + \frac{x^n}{n!} + \cdots$$

which has infinite radius of convergence, so e^x is differentiable for all x with derivative obtained by differentiating the series term by term, i.e., e^x itself.

$\log_e x$ is the inverse function of e^x, so is differentiable for all $x > 0$ by 5.13 with derivative $1/x$.

By definition, $x^a = e^{a \log x}$, so is differentiable for all $x > 0$ by 5.10 with derivative ax^{a-1}.

We can also differentiate $\sin x$ and $\cos x$, and hence all the trigonometric functions. In fact, by definition,

$$\sin x = x - \frac{x^3}{3!} + \frac{x^5}{5!} - \cdots$$

which has infinite radius of convergence, so is differentiable for all x with derivative

$$1 - \frac{x^2}{2!} + \frac{x^4}{4!} - \cdots,$$

i.e. $\cos x$. Similarly $\cos x$ is differentiable for all x with derivative $-\sin x$.

Hence, e.g., $\tan x$ is differentiable for all $x \neq (n + \frac{1}{2})\pi$ with derivative $\sec^2 x$, $\cot x$ is differentiable for all $x \neq n\pi$ with derivative $-\operatorname{cosec}^2 x$, etc.

We must now consider the other applications of the mean value theorem. The first of these concerns uniqueness of primitives.

5.19 Definition

$F(x)$ is a *primitive* (or an *indefinite integral*) of $f(x)$ if $F'(x)=f(x)$. □

5.20 Theorem

If $f(x)$ is differentiable over an interval I, and if $f'(x)=0$ for all $x \in I$, then $f(x)$ must be constant on I.

Proof For any a, $b \in I$ we have

$$\frac{f(b)-f(a)}{b-a}=f'(c)$$

for some c between a and b. Therefore $f'(c)=0$ and hence $f(a)=f(b)$. □

5.21 Corollary

Any two primitives of a given function $f(x)$ must differ by a constant. □

5.22 Application

$$\log(1+x)=x-\frac{x^2}{2}+\frac{x^3}{3}-\frac{x^4}{4}+\cdots$$

for all $|x|<1$.

Proof For all $|x|<1$ we have

$$\frac{1}{1+x}=1-x+x^2-x^3+\cdots$$

since the right-hand side is a geometric series. Now $\log(1+x)$ is a primitive for $1/(1+x)$ and

$$x-\frac{x^2}{2}+\frac{x^3}{3}-\frac{x^4}{4}+\cdots$$

is a primitive for

$$1-x+x^2-x^3+\cdots$$

by 5.18. Therefore

$$\log(1+x) = x - \frac{x^2}{2} + \frac{x^3}{3} - \frac{x^4}{4} + \cdots + C$$

where C is a constant. Putting $x = 0$ clearly gives $C = 0$. □

5.23 Exercise

Prove that

$$\tan^{-1} x = x - \frac{x^3}{3} + \frac{x^5}{5} - \frac{x^7}{7} + \cdots$$

for all $|x| < 1$. *Hint*: Consider primitives for both sides of the expansion

$$\frac{1}{1+x^2} = 1 - x^2 + x^4 - x^6 + \cdots.$$ □

As we remarked at the beginning of the chapter, $f'(x)$ represents the rate of change of $f(x)$. In particular, $f'(x) > 0$ implies $f(x)$ increases, and $f'(x) < 0$ implies $f(x)$ decreases. This apparently self-evident fact is actually quite hard to prove rigorously, at least it would be if we did not have the mean value theorem. With the mean value theorem the proof is spectacularly simple as we now demonstrate.

5.24 Theorem

If $f(x)$ is differentiable over an interval I, and if $f'(x) > 0$ for all $x \in I$, then $f(x)$ is strictly increasing over I. □

Proof Suppose that $a < b$ both $\in I$. Then, by the mean value theorem,

$$\frac{f(b) - f(a)}{b - a} = f'(c)$$

for some $c \in (a, b)$. Therefore $c \in I$ and so $f'(c) > 0$. Hence

$$f(b) - f(a) > 0$$

i.e. $f(a) < f(b)$ as required. □

Similarly, one can prove that, if $f'(x) < 0$ over the interval I, then $f(x)$ decreases over I.

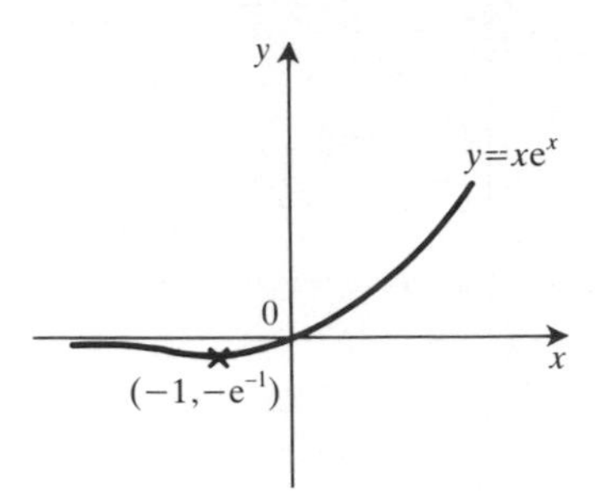

Fig. 5.7

The sceptical reader should attempt to prove 5.24 without the aid of the mean value theorem. It should convince him that the mean value theorem is deeper than it looks, which is really not too surprising when one recalls that it is ultimately based on the Bolzano–Weierstrass theorem (2.38) via the minimax theorem (4.58) and Rolle's theorem (5.15).

Theorem 5.24 can often be used to find maxima and minima of functions. The following example illustrates the method.

5.25 Example

$f(x) = xe^x$.

We have

$$f'(x) = (x+1)e^x$$

so $f'(x) < 0$ if $x < -1$, and $f'(x) > 0$ if $x > -1$. Therefore $f(x)$ decreases strictly over $x < -1$, and increases strictly over $x > -1$. Hence $f(x)$ takes a minimum at $x = -1$ (see Fig. 5.7).

5.26 Exercise

Find maxima and minima of $f(x) = x^2e^x$ and hence sketch the graph.

□

Another application of 5.24 is to the proof of inequalities. By way of illustration we shall give a new proof of Bernoulli's inequality. (See 1.12.)

5.27 Example

For all real $x > -1$, and for all integers $n \geqslant 1$, we have

$$(1+x)^n \geqslant 1 + nx.$$

In fact, let

$$f(x) = (1 + x)^n - 1 - nx.$$

Then

$$f'(x) = n(1 + x)^{n-1} - n\begin{cases} <0 & \text{if } -1 < x < 0 \\ >0 & \text{if } \ x > 0. \end{cases}$$

Therefore $f(x)$ decreases strictly over the interval $-1 < x < 0$, and increases strictly over the interval $x > 0$. It follows that $f(x)$ takes its minimum value at $x = 0$, and hence $f(x) \geqslant f(0)$, i.e.,

$$(1 + x)^n - 1 - nx \geqslant 0,$$

for all $x > -1$, which gives the required result. □

Observe that this method also gives conditions for equality. In fact, for $n \geqslant 2$, we have equality only at $x = 0$.

5.28 Exercises

Prove the following inequalities.

(i) $e^x \geqslant 1 + x$ for all x.
(ii) $\log x \leqslant x - 1$ for all $x > 0$.
(iii) $\sin x < x < \tan x$ for all $0 < x < \frac{1}{2}\pi$.

Investigate conditions for equality in each case. □

In geometrical terms, the mean value theorem says that, between any two points A, B on the graph of a differentiable function, there is a third point C where the tangent is parallel to the chord AB (see Fig. 5.6). This property is enjoyed by plane curves in general, not just those arising as graphs of functions (see Fig. 5.8).

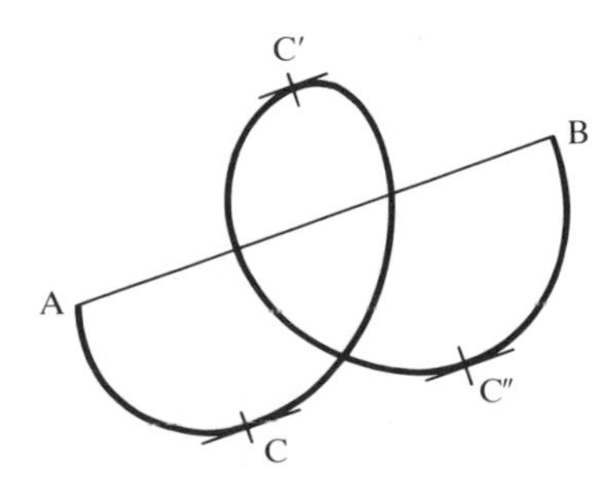

Fig. 5.8

To see what this means arithmetically we represent the curve as the locus of the points having co-ordinates (x, y) where $x = f(t)$, $y = g(t)$ are two functions of a parameter t. Existence of a tangent at every point is ensured by requiring that $f'(t)$, $g'(t)$ both exist and never vanish simultaneously. The slope at the point $(f(t), g(t))$ is then

$$\frac{\mathrm{d}y}{\mathrm{d}x} = \frac{g'(t)}{f'(t)}$$

which is always well defined, though possibly infinite which simply means that the tangent is vertical. The slope of the chord joining the points corresponding to $t = a$, $t = b$ is of course

$$\frac{g(b) - g(a)}{f(b) - f(a)},$$

which again may be infinite with the obvious interpretation.

We arrive at the following theorem.

5.29 Cauchy's mean value theorem

If $f(t)$, $g(t)$ are continuous for all $t \in [a, b]$ and differentiable for all $t \in (a, b)$, and if $f'(t)$, $g'(t)$ do not vanish simultaneously on (a, b), then

$$\frac{g(b) - g(a)}{f(b) - f(a)} = \frac{g'(c)}{f'(c)}$$

for some $c \in (a, b)$.

Proof Consider the function

$$h(t) = (g(b) - g(a))f(t) - (f(b) - f(a))g(t).$$

Clearly, $h(t)$ is continuous on $[a, b]$, differentiable on (a, b) and $h(a) = h(b)$. Therefore, by Rolle's theorem, there exists $c \in (a, b)$ such that $h'(c) = 0$, i.e.,

$$(g(b) - g(a))f'(c) = (f(b) - f(a))g'(c).$$

The condition of $f'(c)$, $g'(c)$ not both being zero ensures that the equation

$$\frac{g(b) - g(a)}{f(b) - f(a)} = \frac{g'(c)}{f'(c)}$$

has a meaning. □

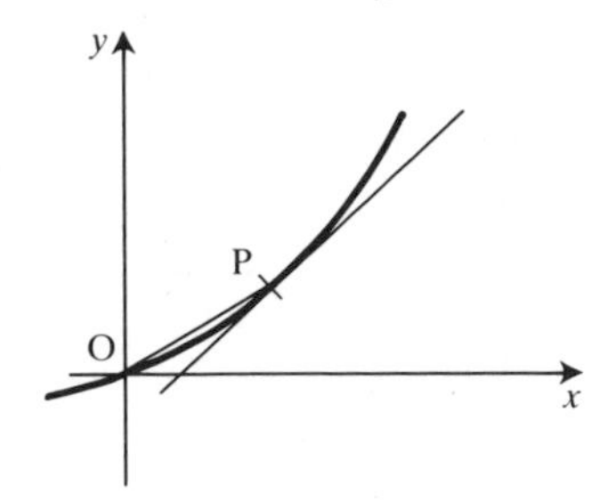

Fig. 5.9

5.30 Application: L'Hôpital's rule

If $f(t)$, $g(t)$ both $\to 0$ as $t \to c$, then

$$\lim_{t\to c}\frac{f(t)}{g(t)} = \lim_{t\to c}\frac{f'(t)}{g'(t)}$$

whenever the second limit exists.

Proof The existence of $\lim_{t\to c} f'(t)/g'(t)$ presupposes that $f'(t)$, $g'(t)$ both exist and $g'(t) \neq 0$ for all t satisfying $0 < |t-c| < \delta$ for some $\delta > 0$. The condition that $\lim_{t\to c} f(t) = \lim_{t\to c} g(t) = 0$ implies that, if we assume $f(c) = g(c) = 0$, then $f(t)$ and $g(t)$ are continuous at c. Therefore, for any t satisfying $0 < |t-c| < \delta$, we can apply Cauchy's mean value theorem to obtain

$$\frac{f(t)}{g(t)} = \frac{f(t)-f(c)}{g(t)-g(c)} = \frac{f'(u)}{g'(u)}$$

for some u between t and c. Now $u \to c$ as $t \to c$ so

$$\lim_{t\to c}\frac{f(t)}{g(t)} = \lim_{u\to c}\frac{f'(u)}{g'(u)} = \lim_{t\to c}\frac{f'(t)}{g'(t)}$$

as required. □

The geometrical interpretation of l'Hôpital's rule is that if a curve passes through the origin O, then the limiting slope of the chord OP is equal to the limiting slope of the tangent at P as $P \to$ O along the curve (see Fig. 5.9).

5.31 Example

$$\lim_{x\to 0}\frac{\log(1+x)}{x} = \lim_{x\to 0}\frac{1}{1+x} = 1.$$ □

5.32 Exercises

Use L'Hôpital's rule to evaluate the following limits.

$$\text{(i)} \lim_{x\to 0} \frac{\sin x}{x} \qquad \text{(ii)} \lim_{x\to \pi} \frac{\sin x}{x-\pi} \qquad \text{(iii)} \lim_{x\to 0} \frac{1-\cos x}{x^2}$$ □

Our final application of the mean value theorem is concerned with the representation of functions by power series.

Suppose we are given a function $f(x)$ and we wish to represent it in thc form

$$f(x) = \sum_0^\infty a_n x^n.$$

If we assume the power series has a positive radius of convergence R then, by 5.18, $f(x)$ must be differentiable and

$$f'(x) = \sum_0^\infty n a_n x^{n-1}$$

for all $|x| < R$. By 5.18 again, $f'(x)$ must be differentiable, i.e. $f(x)$ is *twice* differentiable, and the derivative of $f'(x)$, which we denote by $f''(x)$ and call the *second* derivative of $f(x)$, must be given by

$$f''(x) = \sum_0^\infty n(n-1) a_n x^{n-2}$$

for all $|x| < R$. In fact, $f(x)$ must be differentiable any number of times, and if we denote the kth derivative by $f^{(k)}(x)$, we have

$$f^{(k)}(x) = \sum_0^\infty n(n-1)\cdots(n-k+1) a_n x^{n-k}$$

for all $|x| < R$.

The above power series representations of the successive derivatives of $f(x)$ enable us to determine the coefficients $(a_n)_{n\geqslant 0}$ in terms of $f(x)$. In fact, putting $x = 0$ in the formula for $f^{(k)}(x)$ gives

$$a_k = \frac{1}{k!} f^{(k)}(0)$$

for each $k \geqslant 0$.

5.33 Definition

We say $f(x)$ is *infinitely* differentiable over an interval I if derivatives $f'(x), f''(x), f'''(x), \ldots, f^{(k)}(x), \ldots$ of all orders exist at every $x \in I$. □

5.34 Examples

Any polynomial $p(x)$ is infinitely differentiable for all x. Any rational function $p(x)/q(x)$ is infinitely differentiable except at zeros of the denominator $q(x)$. The functions e^x, $\sin x$, $\cos x$ are infinitely differentiable for all x, and the function $\log x$ is infinitely differentiable for all $x > 0$. □

5.35 Definition

If $f(x)$ is infinitely differentiable at $x = 0$, then

$$a_n = \frac{1}{n!} f^{(n)}(0)$$

is called the nth *Maclaurin coefficient* of $f(x)$, and the series

$$\sum_0^\infty a_n x^n = \sum_0^\infty \frac{1}{n!} f^{(n)}(0) x^n$$

is called the *Maclaurin series* of $f(x)$. □

5.36 Exercises

Find the Maclaurin series of the following functions.

(i) e^x (ii) $\sin x$ (iii) $\cos x$

(iv) $\dfrac{1}{1-x}$ (v) $\log(1+x)$ □

It is natural to expect, or at least hope, that the Maclaurin series of $f(x)$ will converge to $f(x)$ on its interval of convergence. Unfortunately, this is in general not so as the following example demonstrates.

5.37 Example

$$\begin{aligned} f(x) &= e^{-1/x^2} \qquad (x \neq 0), \\ &= 0 \qquad (x = 0). \end{aligned}$$

$f(x)$ is easily shown to be infinitely differentiable for all x with $f^{(n)}(0) = 0$ for all x. Therefore the Maclaurin coefficients of $f(x)$ are all zero, and hence the Maclaurin series of $f(x)$ is trivial, so doesn't converge to $f(x)$ anywhere except at $x = 0$. □

It follows that validity of a Maclaurin expansion has to be verified in each particular case. For many of the important functions of analysis this can be done without too much difficulty. For example, the Maclaurin series of e^x, $\sin x$, $\cos x$ are their defining series, and so validity for all x is immediate. The Maclaurin series of $(1-x)^{-1}$ is the geometric series $\sum_0^\infty x^n$ which we already know converges to $(1-x)^{-1}$ for all $|x|<1$ (see 3.2). The function $\log(1+x)$ has Maclaurin series

$$\sum_1^\infty \frac{(-1)^{n-1}}{n} x^n$$

which converges to $\log(1+x)$ for all $|x|<1$ (see 5.22). We shall show shortly that this expansion is also valid at $x=1$, i.e.

$$\sum_1^\infty \frac{(-1)^{n-1}}{n} = \log 2$$

(see 5.42).

Partial sums of Maclaurin series give polynomial approximations to the functions in question, which can be expected to be fairly accurate, at least for small x. If one requires polynomial approximations for other x, it is more appropriate to use power series of the form

$$\sum_0^\infty a_n(x-a)^n$$

where a is some fixed point $\neq 0$. If

$$f(x) = \sum_0^\infty a_n(x-a)^n,$$

then, as above, the coefficients a_n are given by

$$a_n = \frac{1}{n!} f^{(n)}(a)$$

for all $n \geqslant 0$.

5.38 Definition

If $f(x)$ is infinitely differentiable at $x=a$, then

$$a_n = \frac{1}{n!} f^{(n)}(a)$$

is called the nth *Taylor* coefficient of $f(x)$ at $x = a$, and the series

$$\sum_0^\infty a_n(x-a)^n = \sum_0^\infty \frac{1}{n!} f^{(n)}(a)(x-a)^n$$

is called the *Taylor* series of $f(x)$ at $x = a$. □

5.39 Exercises

Find the Taylor series of the following functions at the points given.

(i) e^x at $x = 2$.

(ii) $\sin x$ at $x = \frac{1}{2}\pi$.

(iii) $\dfrac{1}{1-x}$ at $x = -1$. □

As for Maclaurin series, we cannot expect Taylor series to converge to the function given, though in many important cases they do. (See 5.39.)

5.40 Definition

The Nth *remainder* $R_N(x)$ for the Taylor series of $f(x)$ at $x = a$ is the difference between $f(x)$ and the Nth partial sum of its Taylor series at $x = a$, i.e.,

$$R_N(x) = f(x) - \sum_0^{N-1} a_n(x-a)^n$$

where $a_n = f^{(n)}(a)/n!$. □

Clearly the Taylor series converges to $f(x)$ if and only if $R_N(x) \to 0$ as $N \to \infty$. The following theorem gives a formula for $R_N(x)$ which enables the question of its convergence to zero to be decided in many cases.

5.41 Taylor's theorem with remainder

If $f(x)$ is infinitely differentiable over $[a, x]$, then

$$f(x) = \sum_0^{N-1} \frac{(x-a)^n}{n!} f^{(n)}(a) + \frac{(x-a)^N}{N!} f^{(N)}(c)$$

for some $c \in (a, x)$. □

Observe that case $N=1$ is the mean value theorem. The general case is sometimes called the Nth mean value theorem. An alternative formulation useful in many applications is

$$f(x+h)=\sum_0^{N-1}\frac{h^n}{n!}f^{(n)}(x)+\frac{h^N}{N!}f^{(N)}(x+\theta h)$$

for some $0<\theta<1$, if $f(x)$ is infinitely differentiable over the interval $[x, x+h]$.

Proof of 5.41 Let $g(t)$ be defined by

$$g(t)=\sum_0^{N-1}\frac{(x-t)^n}{n!}f^{(n)}(t)+A\frac{(x-t)^N}{N!}$$

where A is a constant chosen so as to make $g(t)$ satisfy the conditions of Rolle's theorem over $[a, x]$. In fact, A must satisfy the equation

$$f(x)=\sum_0^{N-1}\frac{(x-a)^n}{n!}f^{(n)}(a)+\frac{A(x-a)^N}{N!}.$$

Now the derivative of $g(t)$ is

$$g'(t)=\frac{(x-t)^{N-1}}{(N-1)!}(f^{(N)}(t)-A),$$

so Rolle's theorem gives $A=f^{(N)}(c)$ for some $c\in(a, x)$. The result follows immediately. □

We content ourselves with two applications.

5.42 Theorem

$$\sum_1^{\infty}(-1)^{n-1}/n=\log 2.$$

Proof If $f(x)=\log(1+x)$ and $a=0$, then the Nth remainder $R_N(x)$ is

$$\begin{aligned}R_N(x)&=\frac{x^N}{N!}f^{(N)}(c) \qquad (\text{for some } c\in(0, x))\\&=\frac{x^N}{N!}\frac{(-1)^{N-1}(N-1)!}{(1+c)^N}\\&=\frac{x^N}{N}\frac{(-1)^{N-1}}{(1+c)^N}.\end{aligned}$$

Therefore

$$\begin{aligned}|R_N(1)| &= \frac{1}{N(1+c)^N} \qquad \text{(for some } 0<c<1\text{)}\\ &< \frac{1}{N},\end{aligned}$$

and hence $R_N(1) \to 0$ as $N \to \infty$. □

5.43 Theorem

If $f(x)$ is infinitely differentiable at x and $f'(x)=0$, then $f(x)$ has a maximum at x if $f''(x)<0$, and a minimum if $f''(x)>0$.

Proof The second mean value theorem (5.41 case $N=2$) gives

$$f(x+h) = f(x) + hf'(x) + \tfrac{1}{2}h^2 f''(x+\theta h)$$

for some $0<\theta<1$. If $f''(x)>0$, then there exists $\delta>0$ such that $f''(x+h)>0$ for all $|h|<\delta$, since $f''(x)$ is continuous. (See 5.3.) Therefore for all $|h|<\delta$ we have

$$\begin{aligned}f(x+h) &= f(x) + \tfrac{1}{2}h^2 f''(x+\theta h)\\ &\geqslant f(x)\end{aligned}$$

showing $f(x)$ has a minimum at x. The argument for a maximum is similar.

5.44 Miscellaneous exercises

1. Discuss the differentiability and continuity of the following functions at $x=0$.

(i) $\log|x|$ (ii) $x\log|x|$ (iii) $x^2\log|x|$

(Give each function the value 0 at $x=0$.)

2. Sketch the graphs of the following functions.

(i) e^{-x^2} (ii) xe^{-x^2} (iii) $x^2e^{-x^2}$

3. Prove the following inequalities.

(i) $\sin x > \dfrac{2}{\pi}x \qquad (0<x<\tfrac{1}{2}\pi)$

(ii) $\dfrac{\sin x}{x} > \dfrac{\pi - x}{\pi} \qquad (0<x<\pi)$

(iii) $\dfrac{\sin x}{x} \geqslant \dfrac{\pi^2 - x^2}{\pi^2 + x^2}$ (all real x)

4. Use L'Hôpital's rule (5.30) to find the following limits.

(i) $\lim_{x\to 0} \dfrac{\tan x - x}{x - \sin x}$ (ii) $\lim_{x\to 1} \left(\dfrac{1}{\log x} - \dfrac{1}{x-1}\right)$

(iii) $\lim_{n\to\infty} n(2^{1/n} - 1)$ (iv) $\lim_{n\to\infty} \left(1 + \dfrac{3}{n}\right)^n$

Hint: (for (iii), (iv)) put $x = 1/n$ and let $x \to 0$.

5. Prove that, if $f(x)$ is twice differentiable at x, then

$$f''(x) = \lim_{h\to 0} \frac{f(x) - 2f(x+h) + f(x+2h)}{h^2}.$$

Find the value of

$$\lim_{h\to 0} \frac{f(x) - 3f(x+h) + 3f(x+2h) - f(x+3h)}{h^3}$$

under suitable assumptions on $f(x)$.

Generalize.

6. Given that $f(x)$ is twice differentiable at $x = c$, and that $f'(c) = 0$, $f''(c) > 0$, show that $f(x)$ has a (local) minimum at $x = c$ (cf. 5.43).

What can you say if $f''(c) < 0$? $f''(c) = 0$?

7. Prove that $(x - \alpha)^2$ is a factor of the polynomial $p(x)$ if and only if $p(\alpha) = p'(\alpha) = 0$.

8. Discuss the convergence or otherwise of the series

$$\sum_1^\infty (-1)^{n-1} \frac{\sin \log n}{n}.$$

Hint: Group the terms in pairs and use the mean value theorem. (Observe that the alternating series test is inapplicable here.)

9. Given that $f(0) = 0$, $f''(x) \leqslant 0$ for all $x \geqslant 0$, prove that

$$f(a+b) \leqslant f(a) + f(b)$$

for all $a \geqslant 0$, $b \geqslant 0$. *Hint:* Given $a \leqslant b$, apply the mean value theorem to $f(x)$ over the intervals $[0, a]$, $[b, a+b]$.

Prove the inequality

$$\frac{a+b}{1+a+b} \leqslant \frac{a}{1+a} + \frac{b}{1+b}$$

for all $a \geqslant 0$, $b \geqslant 0$.

10. Prove that, if $f''(x) \geqslant 0$ for all x, then

$$f\left(\frac{a+b}{2}\right) \leqslant \frac{f(a)+f(b)}{2}$$

for all a, b.

Prove conversely that, if the above inequality holds for all a, b, then we must have $f''(x) \geqslant 0$ for all x. *Hint*: (for second part): Use question 5.

11. Show that the Maclaurin series of $(1+x)^\alpha$ is

$$\sum_0^\infty \binom{\alpha}{n} x^n$$

where

$$\binom{\alpha}{n} = \frac{\alpha(\alpha-1)\cdots(\alpha-n+1)}{n!}.$$

Show that it has radius of convergence 1.

Prove that the function

$$f(x) = \sum_0^\infty \binom{\alpha}{n} x^n \qquad (|x|<1)$$

satisfies the differential equation

$$(1+x)f'(x) = \alpha f(x).$$

Deduce that $f(x) = (1+x)^\alpha$ for $|x|<1$.

12. Prove that, for all $x > -1$,

$$(1+x)^\alpha \geqslant 1 + \alpha x,$$

if $\alpha < 0$ or $\alpha > 1$, but

$$(1+x)^\alpha \leqslant 1 + \alpha x$$

if $0 < \alpha < 1$. *Hint* Use 5.41, case $N=2$.

Investigate conditions for equality.

13. Use 5.41 to prove the following inequalities.

(i) $x - \dfrac{x^2}{2} < \log(1+x) < x - \dfrac{x^2}{2} + \dfrac{x^3}{3} \qquad (x>0)$

(ii) $x - \dfrac{x^3}{6} \leqslant \sin x \leqslant x \qquad (x \geqslant 0)$

(iii) $1 - \dfrac{x^2}{2} \leqslant \cos x \leqslant 1 - \dfrac{x^2}{2} + \dfrac{x^4}{24} \qquad$ (all real x)

14. Prove the following assertions.

(i) $\tan^{-1} x + \tan^{-1} y = \tan^{-1} \dfrac{x+y}{1-xy}$.

(ii) $2\tan^{-1}\frac{1}{5} = \tan^{-1}\frac{5}{12}$.

(iii) $4\tan^{-1}\frac{1}{5} = \tan^{-1}\frac{120}{119}$.

(iv) $\tan^{-1}\frac{120}{119} - \tan^{-1} 1 = \tan^{-1}\frac{1}{239}$.

(v) $\frac{1}{4}\pi = 4\tan^{-1}\frac{1}{5} - \tan^{-1}\frac{1}{239}$.

(v) is known as Machin's formula, and has been used to calculate π to vast numbers of decimal places by using the expansion

$$\tan^{-1} x = x - \frac{x^3}{3} + \frac{x^5}{5} - \frac{x^7}{7} + \cdots.$$

(See 5.23.)

15. Prove that, for any integer $N \geqslant 1$,

$$\mathrm{e} = \sum_{0}^{N-1} \frac{1}{n!} + \frac{\mathrm{e}^{u_N}}{N!}$$

for some $0 < u_N < 1$. Deduce that e is irrational. *Hint:* If $\mathrm{e} = p/q$ where p, q are positive integers, then e^{u_N}/N is an integer for all $N > q$.

6
Integral calculus

There are two essentially different ways of looking at integration. One can either regard it as the inverse process of differentiation, or one can regard it as a kind of continuous summation.

The first view gives rise to the 'indefinite integral' or 'primitive' of a function $f(x)$ defined as any function $F(x)$ whose derivative $F'(x)=f(x)$ (see 5.19). The second view gives rise to the 'definite integral' of a function $f(x)$ which is defined as the limiting value of sums of the form $\sum f(x)\,\mathrm{d}x$ as $\mathrm{d}x \to 0$. The definite integral $\int_a^b f(x)\,\mathrm{d}x$ admits a geometrical interpretation as the area under the graph of $y=f(x)$ between the limits a and b (see Fig. 6.1).

The two kinds of integral are of course intimately related. The definite integral is expressible in terms of the indefinite integral as

$$\int_a^b f(x)\,\mathrm{d}x = F(b) - F(a),$$

where $F(x)$ is any indefinite integral of $f(x)$. On the other hand, the definite integral of $f(x)$ can be used to define an indefinite integral $F(x)$ of $f(x)$ by writing

$$F(x) = \int_a^x f(t)\,\mathrm{d}t.$$

These two results are collectively known as the fundamental theorem of calculus, and it is the main object of this chapter to give a rigorous proof this theorem for continuous functions.

The difficult problem is to give a rigorous definition of $\lim_{\mathrm{d}x\to 0} \sum f(x)\,\mathrm{d}x$ since this is really rather a sophisticated kind of limit. We shall need a more adequate notation for $\sum f(x)\,\mathrm{d}x$ which gives a more detailed description of how the interval $[a, b]$ is to be subdivided into small pieces $\mathrm{d}x$. We shall then need to find a precise way of describing how $\sum f(x)\,\mathrm{d}x$ tends to a limit as the small pieces $\mathrm{d}x$ all tend to zero simultaneously. The discussion will be motivated by thinking of the definite integral as the area under the graph.

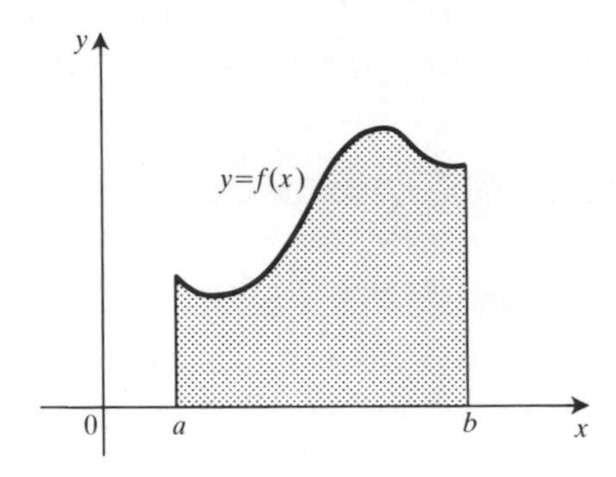

Fig. 6.1

6.1 Definitions

A *dissection* D of the interval $[a, b]$ is any finite set of points $x_0, x_1, x_2, \ldots, x_N$ such that

$$a = x_0 < x_1 < x_2 < \cdots < x_N = b.$$

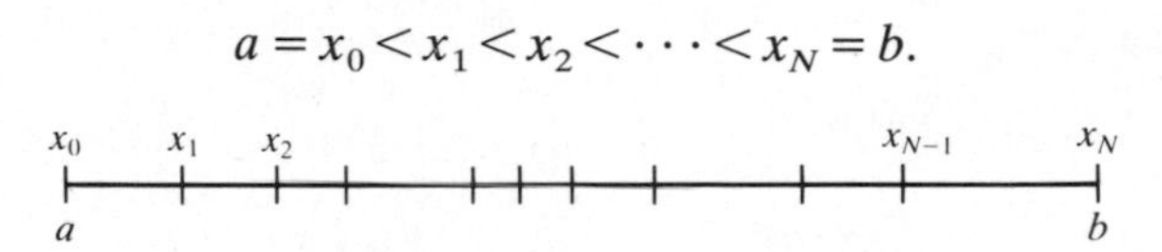

Given any bounded function $f(x)$ over the interval $[a, b]$, we define its *upper sum* and *lower sum* for the dissection D, denoted by U_D and L_D, to be

$$U_D = \sum_1^N M_n(x_n - x_{n-1}),$$

$$L_D = \sum_1^N m_n(x_n - x_{n-1})$$

where

$$M_n = \sup_{x_{n-1} \leqslant x \leqslant x_n} f(x),$$

$$m_n = \inf_{x_{n-1} \leqslant x \leqslant x_n} f(x)$$

(see Fig. 6.2). □

Observe that U_D, L_D approximate the area under the curve from above and below by rectilinear areas.

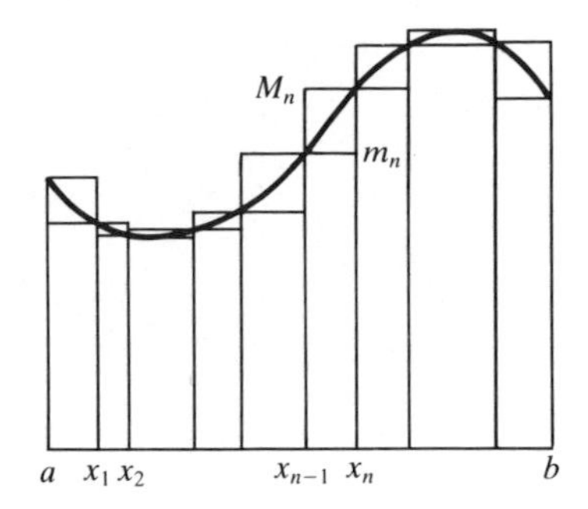

Fig. 6.2

6.2 Exercise

Compute U_D, L_D for $f(x)=x$ over the interval $[0, 1]$ corresponding to the dissection D given by

$$0<\frac{1}{N}<\frac{2}{N}<\cdots<\frac{N-1}{N}<1.$$ □

6.3 Lemma

$$m(b-a)\leqslant L_D\leqslant U_D\leqslant M(b-a)$$

where

$$M=\sup_{a\leqslant x\leqslant b} f(x),$$
$$m=\inf_{a\leqslant x\leqslant b} f(x).$$ □

Proof Clearly

$$m\leqslant m_n\leqslant M_n\leqslant M$$

for all n, and therefore

$$m\sum_1^N (x_n-x_{n-1})\leqslant\sum_1^N m_n(x_n-x_{n-1})\leqslant\sum_1^N M_n(x_n-x_{n-1})$$
$$\leqslant M\sum_1^N (x_n-x_{n-1}),$$

which is the result. □

6.4 Definition

We say D' is a *refinement* of D, and we write $D'>D$, if D' can be obtained from D by adding further points. □

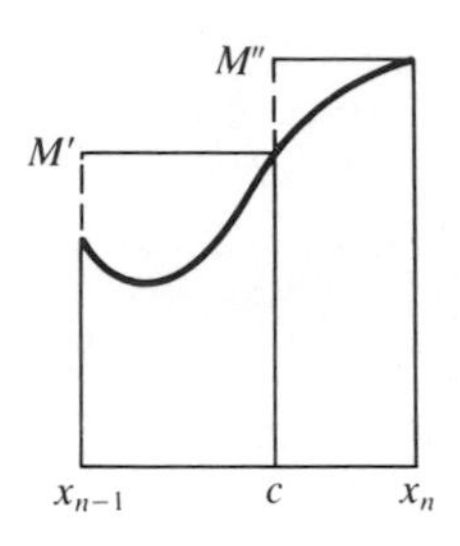

Fig. 6.3

For example, if D_N is the dissection of $[a, b]$ into N equal subintervals, then

$$D_1 < D_2 < D_4 < \cdots < D_{2^N} < \cdots.$$

6.5 Lemma

If $D' > D$, then $U_{D'} \leqslant U_D$ and $L_{D'} \geqslant L_D$. □

Proof Suppose D is

$$a = x_0 < x_1 < x_2 < \cdots < x_N = b$$

and that D' adds one further point c between x_{n-1} and x_n. Let

$$M' = \sup_{x_{n-1} \leqslant x \leqslant c} f(x),$$
$$M'' = \sup_{c \leqslant x \leqslant x_n} f(x)$$

(see Fig. 6.3). Then clearly $M' \leqslant M_n$, $M'' \leqslant M_n$, and therefore

$$\begin{aligned} U_{D'} &= \sum_{r \neq n} M_r(x_r - x_{r-1}) + M'(c - x_{n-1}) + M''(x_n - c) \\ &\leqslant \sum_{r \neq n} M_r(x_r - x_{r-1}) + M_n(c - x_{n-1}) + M_n(x_n - c) \\ &= U_D. \end{aligned}$$

Hence clearly $U_{D'} \leqslant U_D$ for any number of added points.

The proof of $L_{D'} \geqslant L_D$ is similar. □

6.6 Definition

Given any two dissections D, D' we define their *common refinement*, denoted by $D \cup D'$, to be the dissection obtained by combining all the points of D with all the points of D'. □

6.7 Lemma

For any D, D' we have $L_D \leqslant U_{D'}$. □

Proof Lemmas 6.3 and 6.5 give

$$L_D \leqslant L_{D\cup D'} \leqslant U_{D\cup D'} \leqslant U_{D'}.$$ □

6.8 Definition

The *upper and lower integrals* of bounded $f(x)$ over $[a, b]$, denoted by U, L, are defined to be

$$U = \inf_D U_D,$$

$$L = \sup_D L_D.$$ □

Observe that U is the limiting value of rectilinear areas exterior to the area under the graph, L is the limiting value of interior rectilinear areas.

6.9 Lemma

$$m(b-a) \leqslant L \leqslant U \leqslant M(b-a).$$

Proof follows from 6.3 and 6.7. □

6.10 Definition

We say bounded $f(x)$ is *integrable* over $[a, b]$ if $L = U$ and when this occurs we define the *definite integral* of $f(x)$ over $[a, b]$, denoted by $\int_a^b f(x)\,\mathrm{d}x$, to be the common value of L and U. □

We have to introduce a notion of 'integrability' because it can happen that $L \neq U$ as we shall shortly see. It turns out that $L = U$ for a wide enough class of functions to make the above definition useful for our present purposes. In particular, we can show monotonic functions and continuous functions are integrable over any interval. Non-integrable functions tend to be intellectual curiosities, but they are none the less useful for demonstrating the limitations of this definition of the definite integral.

6.11 Examples

(i) $f(x) \equiv C$ constant.

For any dissection D of any interval $[a, b]$ we have

$$M_n = m_n = C$$

for all n, and therefore

$$U_D = L_D = C(b-a).$$

Hence $f(x)$ is integrable over $[a, b]$ and

$$\int_a^b f(x)\,dx = \int_a^b C\,dx = C(b-a).$$

(ii) $f(x) = 1$ if $x = a_1, a_2, \ldots, a_n$,
$= 0$ otherwise.

For any dissection D of any interval $[a, b]$ we clearly have $L_D = 0$. Therefore $L = 0$. By choosing D to include the points $a_1 \pm \varepsilon, a_2 \pm \varepsilon, \ldots, a_n \pm \varepsilon$ we clearly have

$$U_D \leqslant 2n\varepsilon$$

for any $\varepsilon > 0$. Hence $U = 0$ and so $f(x)$ is integrable over $[a, b]$ and $\int_a^b f(x)\,dx = 0$.

(iii) *Dirichlet's function* $f(x) = 1$ if x rational,
$= 0$ otherwise.

For any dissection D of any interval $[a, b]$ we have $M_n = 1$, $m_n = 0$ for all n. (See 1.31.) Therefore $U_D = b - a$, $L_D = 0$. Hence

$$U = b - a > 0 = L$$

and so $f(x)$ is not integrable over $[a, b]$. □

6.12 Exercise

Let $f(x) = x$, $[a, b] = [0, 1]$. Let D_N be the dissection of $[0, 1]$ into N equal subintervals. Show U_{D_N}, L_{D_N} both $\to \frac{1}{2}$ as $N \to \infty$ (cf. 6.2).

Deduce $f(x) = x$ is integrable over $[0, 1]$ and that $\int_0^1 x\,dx = \frac{1}{2}$. *Hint:* Observe that $L_{D_N} \leqslant L \leqslant U \leqslant U_{D_N}$ and let $N \to \infty$. □

6.13 Theorem: Riemann's condition

Bounded $f(x)$ is integrable over $[a, b]$ if and only if, given any $\varepsilon > 0$, there exists a dissection D of $[a, b]$ such that

$$U_D - L_D < \varepsilon.$$

Proof Suppose $f(x)$ is bounded integrable over $[a, b]$ and $\varepsilon > 0$ is given. Then we can find dissections D, D' such that

$$L_D > L - \tfrac{1}{2}\varepsilon,$$

$$U_{D'} < U + \tfrac{1}{2}\varepsilon. \qquad \text{(1.34, question 6)}$$

Therefore

$$\begin{aligned} U_{D \cup D'} - L_{D \cup D'} &\leqslant U_{D'} - L_D \qquad \text{(see 6.5)} \\ &< (U + \tfrac{1}{2}\varepsilon) - (L - \tfrac{1}{2}\varepsilon) \\ &= \varepsilon \end{aligned}$$

since $L = U$.

Suppose on the other hand that Riemann's condition is satisfied. Let $\varepsilon > 0$ be given, and let D be chosen in accordance with the condition. Then we have

$$\begin{aligned} 0 &\leqslant U - L \\ &\leqslant U_D - L_D \\ &< \varepsilon; \end{aligned}$$

therefore

$$0 \leqslant U - L < \varepsilon,$$

and hence $L = U$, since ε is arbitrarily small. □

We can use 6.13 to show monotonic functions and continuous functions are integrable as follows.

6.14 Theorem

If $f(x)$ increases over $[a, b]$, then $f(x)$ is integrable over $[a, b]$. □

Proof Let D be the dissection

$$a = x_0 < x_1 < x_2 < \cdots < x_N = b$$

of $[a, b]$. Then we have $M_n = f(x_n)$, $m_n = f(x_{n-1})$ for all n, and

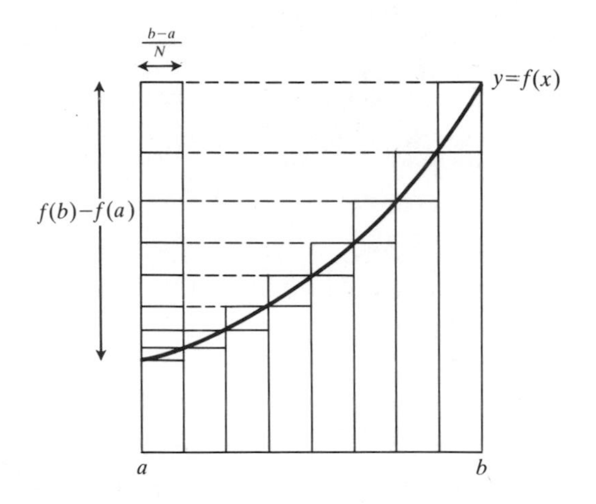

Fig. 6.4

hence

$$\begin{aligned} U_D - L_D &= \sum_1^N (M_n - m_n)(x_n - x_{n-1}) \\ &= \sum_1^N (f(x_n) - f(x_{n-1}))(x_n - x_{n-1}) \\ &= \frac{b-a}{N} \sum_1^N (f(x_n) - f(x_{n-1})), \end{aligned}$$

if we choose D such that $x_n - x_{n-1} = (b-a)/N$ for all n,

$$= \frac{(b-a)(f(b)-f(a))}{N}$$

which is arbitrarily small if N is chosen large enough. □

Observe that $U_D - L_D$ is the total area of the small boxes covering the graph of $f(x)$. These boxes can all be slid horizontally to the left to form a pile of height $f(b) - f(a)$ and width $(b-a)/N$ (see Fig. 6.4).

6.15 Theorem

If $f(x)$ is continuous over $[a, b]$, then $f(x)$ is integrable over $[a, b]$.

Proof Observe firstly that $f(x)$ is bounded over $[a, b]$ by the minimax theorem (4.58).

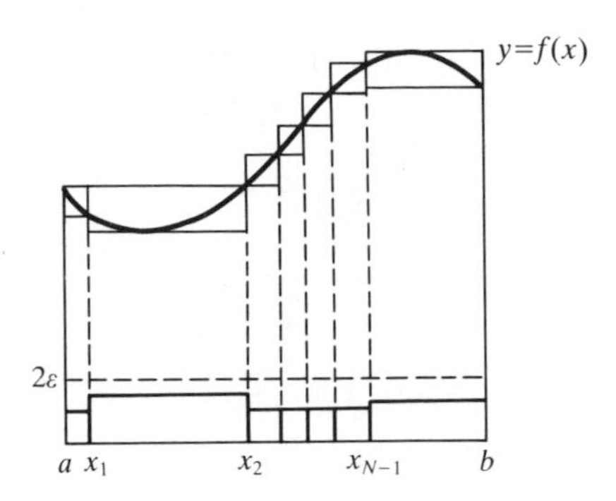

Fig. 6.5

Suppose $\varepsilon > 0$ is given. Construct a dissection D of $[a, b]$ as follows. Let x_1 be the first point in $[a, b]$ such that

$$|f(x_1) - f(a)| = \varepsilon.$$

(See 4.59, question 10.) If no such point exists take $x_1 = b$. If $x_1 < b$, let x_2 be the first point in $[x_1, b]$ such that

$$|f(x_2) - f(x_1)| = \varepsilon.$$

If no such point exists take $x_2 = b$. Define $x_3, x_4, \ldots$ similarly. We must have $x_N = b$ for some N. This is because, if not, then we would have an infinite sequence $(x_n)_{n \geq 1}$ such that

$$x_1 < x_2 < \cdots < x_n < \cdots < b$$

and

$$|f(x_n) - f(x_{n+1})| = \varepsilon$$

for all n. Therefore $(x_n)_{n \geq 1}$, being increasing and bounded above, would converge to a limit x, say. Hence, by continuity, $f(x_n)$, $f(x_{n+1})$ would both converge to $f(x)$, giving a contradiction.

This dissection D has the property that for every n

$$M_n - m_n < 2\varepsilon.$$

Therefore

$$\begin{aligned} U_D - L_D &= \sum_1^N (M_n - m_n)(x_n - x_{n-1}) \\ &< 2\varepsilon \sum_1^N (x_n - x_{n-1}) \\ &= 2\varepsilon(b - a) \end{aligned}$$

which is arbitrarily small. Hence $f(x)$ is integrable over $[a, b]$ by 6.13. □

Observe that, as in 6.14, $U_D - L_D$ is represented by the total area of the boxes covering the graph of $f(x)$. By sliding these all down to the x-axis we obtain an area included in a rectangle of height 2ε and breadth $b-a$ (see Fig. 6.5).

6.16 Exercise

Which functions are integrable over the intervals stated?

(i) $\operatorname{sgn} x$ over $[-1, 1]$.

(ii) $|x|$ over $[-1, 1]$.

(iii) $f(x) = 1$ if $1 < x < 2$,

$= 0$ otherwise,

over $[0, 3]$. □

The reader will have observed that, while we have given a precise definition of the definite integral $\int_a^b f(x)\,dx$ for certain functions $f(x)$, we haven't quite related it to $\lim_{dx\to 0} \Sigma f(x)\,dx$. This will be our next objective.

6.17 Definition

If D is the dissection

$$a = x_0 < x_1 < x_2 < \cdots < x_N = b$$

of $[a, b]$, then we define the *modulus* of D, denoted by $|D|$, to be

$$|D| = \max_{1 \leq n \leq N} (x_n - x_{n-1}).$$ □

6.18 Lemma

If D, D' are two dissections such that $|D| < \delta$ and D' has N' interior points, then

$$U_{D \cup D'} > U_D - \delta N'(M - m)$$

where, as usual,

$$M = \sup_{a \leq x \leq b} f(x),$$

$$m = \inf_{a \leq x \leq b} f(x).$$ □

Proof If D' has one interior point c which occurs in the nth subinterval

(x_{n-1}, x_n) of D, then

$$\begin{aligned} U_D - U_{D\cup D'} &= M_n(x_n - x_{n-1}) - M'(c - x_{n-1}) - M''(x_n - c) \\ &\qquad\qquad\qquad\qquad \text{(in the notation of 6.5)} \\ &\leqslant M(x_n - x_{n-1}) - m(c - x_{n-1}) - m(x_n - c) \\ &= (M - m)(x_n - x_{n-1}) \\ &< (M - m)\delta. \end{aligned}$$

Therefore, if D' has N' interior points,

$$U_D - U_{D\cup D'} < N'(M - m)\delta$$

which gives the result. □

6.19 Theorem

If $f(x)$ is bounded integrable over $[a, b]$, then U_D, L_D both $\rightarrow \int_a^b f(x)\,dx$ as $|D| \rightarrow 0$.

Proof Suppose $\varepsilon > 0$ is given. Choose D' such that

$$U_{D'} < U + \tfrac{1}{2}\varepsilon. \qquad \text{(1.34, question 6)}$$

and suppose D' has N' interior points. Choose

$$\delta = \frac{\varepsilon}{2N'(M - m)}.$$

Then, for any D with $|D| < \delta$, we have

$$\begin{aligned} U_D &< U_{D\cup D'} + \delta N'(M - m) \qquad (6.18) \\ &= U_{D\cup D'} + \tfrac{1}{2}\varepsilon \\ &\leqslant U_{D'} + \tfrac{1}{2}\varepsilon \qquad (6.5) \\ &< U + \varepsilon. \end{aligned}$$

Hence

$$U \leqslant U_D < U + \varepsilon$$

for all $|D| < \delta$, which shows that $U_D \rightarrow U$ as $|D| \rightarrow 0$.

Similarly $L_D \rightarrow L$ as $|D| \rightarrow 0$. □

6.20 Definition

A *Riemann sum* Σ_D for bounded $f(x)$ corresponding to a dissection D of $[a, b]$ is any sum of the form

$$\Sigma_D = \sum_1^N f(c_n)(x_n - x_{n-1})$$

where $c_n \in [x_{n-1}, x_n]$ for each $1 \leqslant n \leqslant N$. □

If $f(x)$ is continuous, then U_D, L_D are Riemann sums (by the minimax theorem 4.58), but in general they may not be. Observe that the notation for a Riemann sum should include mention of the points $c_n \in [x_{n-1}, x_n]$. We have chosen to sacrifice a certain amount of precision in the interests of clarity.

6.21 Darboux' theorem

If bounded $f(x)$ is integrable over $[a, b]$, then the Riemann sums $\Sigma_D \to \int_a^b f(x)\,\mathrm{d}x$ as $|D| \to 0$. □

This is the promised rigorization of '$\Sigma f(x)\,\mathrm{d}x \to \int_a^b f(x)\,\mathrm{d}x$ as $\mathrm{d}x \to 0$'.

Proof For any Riemann sum Σ_D corresponding to any dissection D we clearly have

$$L_D \leqslant \Sigma_D \leqslant U_D,$$

so the result follows by the sandwich principle (2.16) and 6.19. □

We can now make a start on the fundamental theorem of calculus.

6.22 Theorem

If $f'(x)$ exists and is continuous on $[a, b]$, then

$$\int_a^b f'(x)\,\mathrm{d}x = f(b) - f(a).$$

Proof For any dissection D by

$$a = x_0 < x_1 < \cdots < x_N = b$$

of $[a, b]$, we have

$$f(b)-f(a)=\sum_1^N (f(x_n)-f(x_{n-1}))$$
$$=\sum_1^N f'(c_n)(x_n-x_{n-1}),$$

for some $c_n \in (x_{n-1}, x_n)$ by the mean value theorem (5.16),

$$\to \int_a^b f'(x)\,\mathrm{d}x$$

as $|D| \to 0$ by Darboux' theorem (6.21). Hence the result. □

Theorem 6.22 enables a large class of definite integrals to be evaluated. For example,

$$\int_a^b \mathrm{e}^x\,\mathrm{d}x = \mathrm{e}^b - \mathrm{e}^a,$$
$$\int_a^b \cos x\,\mathrm{d}x = \sin b - \sin a.$$

We shall have to postpone the proof of the other fundamental theorem of calculus—that the derivative of

$$F(x)=\int_a^x f(t)\,\mathrm{d}t$$

is $f(x)$—until we have established some basic properties of the definite integral.

6.23 Linearity of the integral

If $f(x)$, $g(x)$ are continuous over $[a, b]$, and if C is a constant, then

(i) $\displaystyle\int_a^b (f(x)+g(x))\,\mathrm{d}x = \int_a^b f(x)\,\mathrm{d}x + \int_a^b g(x)\,\mathrm{d}x,$

(ii) $\displaystyle\int_a^b Cf(x)\,\mathrm{d}x = C\int_a^b f(x)\,\mathrm{d}x.$

Proof For any dissection D of $[a, b]$ by

$$a=x_0<x_1<x_2<\cdots<x_N=b$$

and $c_n \in [x_{n-1}, x_n]$ for each $1 \leqslant n \leqslant N$, we have

$$\sum_1^N (f(c_n)+g(c_n))(x_n - x_{n-1}) = \sum_1^N f(c_n)(x_n - x_{n-1}) + \sum_1^N g(c_n)(x_n - x_{n-1}),$$

$$\sum_1^N Cf(c_n)(x_n - x_{n-1}) = C\sum_1^N f(c_n)(x_n - x_{n-1}),$$

so the result follows from 6.21 by letting $|D| \to 0$. □

6.24 Integration of inequalities

If $f(x) \leqslant g(x)$ are continuous for all $x \in [a, b]$, then

$$\int_a^b f(x)\,\mathrm{d}x \leqslant \int_a^b g(x)\,\mathrm{d}x.$$

Proof For any dissection D by

$$a = x_0 < x_1 < \cdots < x_N = b$$

and c_n from each subinterval $[x_{n-1}, x_n]$, we have

$$\sum_1^N f(c_n)(x_n - x_{n-1}) \leqslant \sum_1^N g(c_n)(x_n - x_{n-1}),$$

so the result follows by letting $|D| \to 0$. □

6.25 Modular inequality for integrals

If $f(x)$ is continuous on $[a, b]$, then

$$\left|\int_a^b f(x)\,\mathrm{d}x\right| \leqslant \int_a^b |f(x)|\,\mathrm{d}x.$$

Proof For any

$$a = x_0 < x_1 < \cdots < x_N = b$$

and $c_n \in [x_{n-1}, x_n]$, we have

$$\left|\sum_1^N f(c_n)(x_n - x_{n-1})\right| \leqslant \sum_1^N |f(c_n)|(x_n - x_{n-1}),$$

so the result follows by 6.21. □

6.26 Interval additivity

If $a<b<c$ and $f(x)$ is continuous over $[a, c]$, then

$$\int_a^c f(x)\,\mathrm{d}x = \int_a^b f(x)\,\mathrm{d}x + \int_b^c f(x)\,\mathrm{d}x.$$

Proof Any dissection D of $[a, c]$ which includes the point b, say

$$a = x_0 < x_1 < \cdots < x_n = b < x_{n+1} < \cdots < x_N = c,$$

generates dissections D', D'' of $[a, b]$, $[b, c]$, namely

$$a = x_0 < x_1 < \cdots < x_n = b,$$

$$b = x_n < x_{n+1} < \cdots < x_N = c.$$

Also, for any $c_r \in [x_{r-1}, x_r]$ for each $1 \leqslant r \leqslant N$, we have

$$\sum_{r=1}^{N} f(c_r)(x_r - x_{r-1}) = \sum_{r=1}^{n} f(c_r)(x_r - x_{r-1}) + \sum_{r=n+1}^{N} f(c_r)(x_r - x_{r-1}),$$

that is

$$\Sigma_D = \Sigma_{D'} + \Sigma_{D''},$$

and if we let $|D| \to 0$, then clearly $|D'|$, $|D''|$ both $\to 0$ too, so the result follows from 6.21. □

6.27 Definition

If $a<b$ and $f(x)$ is continuous over $[a, b]$ we define $\int_b^a f(x)\,\mathrm{d}x$ to be $-\int_a^b f(x)\,\mathrm{d}x$. □

6.28 Theorem

If $f(x)$ is continuous over $[a, b]$ and x, y, z are any three points in $[a, b]$, then

$$\int_x^y f(t)\,\mathrm{d}t + \int_y^z f(t)\,\mathrm{d}t = \int_x^z f(t)\,\mathrm{d}t.$$ □

Proof is easily achieved by enumerating the various orderings of x, y, z along the line. □

We can now complete our discussion of the fundamental theorem of calculus.

6.29 Theorem

If $f(t)$ is continuous over $[a, b]$ and

$$F(x) = \int_a^x f(t)\,\mathrm{d}t,$$

then $F(x)$ is differentiable over $[a, b]$ and its derivative is $F'(x) = f(x)$. □

Proof If $x \in [a, b]$ and $\varepsilon > 0$ are given, then we can choose $\delta > 0$ such that

$$|f(t) - f(x)| < \varepsilon$$

for all $t \in [a, b]$ satisfying $|t - x| < \delta$. Therefore, for any $0 < h < \delta$, we have

$$\begin{aligned}
\left|\frac{F(x+h) - F(x)}{h} - f(x)\right| &= \left|\frac{\int_a^{x+h} f(t)\,\mathrm{d}t - \int_a^x f(t)\,\mathrm{d}t}{h} - f(x)\right| \\
&= \left|\frac{1}{h}\int_x^{x+h} f(t)\,\mathrm{d}t - f(x)\right| \qquad (6.28) \\
&= \left|\frac{1}{h}\int_x^{x+h} (f(t) - f(x))\,\mathrm{d}t\right| \qquad (6.11(\mathrm{i})) \\
&\leqslant \frac{1}{h}\int_x^{x+h} |f(t) - f(x)|\,\mathrm{d}t \qquad (6.25) \\
&\leqslant \frac{1}{h}\int_x^{x+h} \varepsilon\,\mathrm{d}t \qquad (6.24) \\
&= \varepsilon.
\end{aligned}$$

Therefore $F'_+(x) = f(x)$, and $F'_-(x) = f(x)$ is proved similarly (see 5.5). □

6.30 Corollary

Any continuous function $f(x)$ on the interval $[a, b]$ has a primitive on $[a, b]$.

Proof By 6.29 the function

$$F(x) = \int_a^x f(t)\,\mathrm{d}t$$

is a primitive of $f(t)$. □

In applications of integral calculus the problem invariably boils down to finding primitives. In all but a few cases this problem is quite difficult. Various techniques are available for finding primitives of apparently intractable functions. Most important among these are integration by parts and by substitution which we shall now consider.

6.31 Integration by parts

If $f'(x)$, $g'(x)$ exist and are continuous over $[a, b]$, then

$$\int_a^b f'(x)g(x)\,\mathrm{d}x = f(b)g(b) - f(a)g(a) - \int_a^b f(x)g'(x)\,\mathrm{d}x.$$

Proof If $h(x) = f(x)g(x)$ then, by 6.22,

$$\int_a^b h'(x)\,\mathrm{d}x = h(b) - h(a).$$

Therefore

$$\int_a^b (f'(x)g(x) + f(x)g'(x))\,\mathrm{d}x = f(b)g(b) - f(a)g(a), \qquad (5.6)$$

from which the result follows. □

6.32 Worked example

Evaluate $\int_0^1 x\mathrm{e}^x\,\mathrm{d}x$.

Let $f'(x) = \mathrm{e}^x$, $g(x) = x$. Then we can take $f(x) = \mathrm{e}^x$ and we have $g'(x) = 1$. Therefore, by 6.31,

$$\begin{aligned}\int_0^1 x\mathrm{e}^x\,\mathrm{d}x &= x\mathrm{e}^x\big|_0^1 - \int_0^1 \mathrm{e}^x\,\mathrm{d}x \\ &= \mathrm{e} - \mathrm{e}^x\big|_0^1 \\ &= \mathrm{e} - (\mathrm{e} - 1) \\ &= 1.\end{aligned}$$

□

Observe that we are using the notation

$$f(x)\big|_a^b = f(b) - f(a).$$

6.33 Exercises

Evaluate the following integrals.

(i) $\int_0^{\pi} x \cos x \, \mathrm{d}x$ (ii) $\int_0^1 x^2 \mathrm{e}^x \, \mathrm{d}x$

(iii) $\int_0^{\pi} \mathrm{e}^x \sin x \, \mathrm{d}x$ (iv) $\int_1^2 \log x \, \mathrm{d}x$ □

6.34 Integration by substitution

If $f(x)$ is continuous over $[a, b]$ and if $g'(t)$ exists and is continuous over $[\alpha, \beta]$, where $g(\alpha) = a$, $g(\beta) = b$, then

$$\int_a^b f(x) \, \mathrm{d}x = \int_\alpha^\beta f(g(t))g'(t) \, \mathrm{d}t.$$

Proof Let $F(x)$ be a primitive of $f(x)$ (see 6.30). If $h(t) = F(g(t))$ then, by 6.22,

$$\int_\alpha^\beta h'(t) \, \mathrm{d}t = h(\beta) - h(\alpha).$$

Therefore

$$\begin{aligned} \int_\alpha^\beta F'(g(t))g'(t) \, \mathrm{d}t &= F(g(\beta)) - F(g(\alpha)) \quad (5.10) \\ &= F(b) - F(a) \\ &= \int_a^b F'(x) \, \mathrm{d}x, \qquad (6.22) \end{aligned}$$

that is

$$\int_\alpha^\beta f(g(t))g'(t) \, \mathrm{d}t = \int_a^b f(x) \, \mathrm{d}x,$$

which is the required result. □

6.35 Worked examples

(i) Evaluate $\int_0^1 \frac{\mathrm{d}x}{1 + x^2}$.

Let $f(x) = 1/(1 + x^2)$, $g(t) = \tan t$. Then $g'(t) = \sec^2 t$ and therefore, by 6.34,

$$\begin{aligned} \int_0^1 \frac{\mathrm{d}x}{1 + x^2} &= \int_0^{\pi/4} \frac{1}{1 + \tan^2 t} \sec^2 t \, \mathrm{d}t \\ &= \int_0^{\pi/4} \mathrm{d}t \\ &= \pi/4. \end{aligned}$$

(ii) Evaluate $\int_0^1 e^{2t}\,dt$.

Let $f(x)=e^x$, $g(t)=2t$. Then $g'(t)=2$ and therefore, by 6.34,

$$\int_0^1 e^{2t}\,.\,2\,dt=\int_0^2 e^x\,dx,$$

and hence

$$\int_0^1 e^{2t}\,dt=\tfrac{1}{2}\int_0^2 e^x\,dx$$
$$=\tfrac{1}{2}e^x\big|_0^2$$
$$=\tfrac{1}{2}(e^2-1).$$

□

6.36 Exercises

Evaluate the following integrals.

(i) $\displaystyle\int_0^{\frac{1}{2}} \frac{dx}{\sqrt{(1-x^2)}}$ (ii) $\displaystyle\int_0^1 \frac{dx}{1+\sqrt{x}}$

(iii) $\displaystyle\int_0^{\frac{1}{2}\pi} \sin\,(t+\tfrac{1}{2}\pi)\,dt$ (iv) $\displaystyle\int_0^1 \frac{dt}{1+\sqrt{t}}$

Hints: Make the substitutions $x=\sin t$, t^2, $t+\frac{1}{2}\pi$, $\sqrt{t}$. □

There are certain definite integrals which cannot be evaluated in closed form because their integrands (functions to be integrated) have no primitives expressible in terms of the elementary functions. Examples of such integrals are $\int_0^1 e^{x^2}\,dx$, $\int_0^{\pi} \sin x/x\,dx$. To evaluate these integrals one would have to resort to numerical methods, and be content with an approximate answer. Of course, e.g., $f(x)=e^{x^2}$ does have a primitive, namely

$$F(x)=\int_0^x e^{t^2}\,dt,$$

but $F(x)$ cannot be expressed in a simpler form.

6.37 Miscellaneous exercises

1. Discuss the integrability over $[0, 1]$ of the following functions.

$$\text{(i) } \sin\frac{1}{x} \qquad \text{(ii) } x\sin\frac{1}{x}$$

(Set each equal to 0 at $x=0$.)

2. Prove the following inequalities.

(i) $\frac{1}{2}\pi \leqslant \int_0^\pi \frac{\sin x}{x}\,\mathrm{d}x \leqslant \pi.$

(ii) $\frac{2}{3} \leqslant \int_0^1 \mathrm{e}^{-x^2}\,\mathrm{d}x \leqslant \frac{21}{30}.$

Hints: Integrate the inequalities

(i) $\frac{\pi - x}{\pi} \leqslant \frac{\sin x}{x} \leqslant 1,$ (5.44, question 3)

(ii) $1 - x^2 \leqslant \mathrm{e}^{-x^2} \leqslant 1 - x^2 + \frac{x^4}{2}.$

3. Prove that if $f'(x)$ exists and is continuous over $[a, b]$, then

$$\int_a^b f(x) \sin nx\,\mathrm{d}x \to 0$$

as $n \to \infty$. *Hint:* Integrate by parts.

4. Prove that, if $f(t)$ is bounded integrable over $[a, x]$ for every $x \in [a, b]$, then

$$F(x) = \int_a^x f(t)\,\mathrm{d}t$$

is continuous, but may fail to be differentiable if $f(t)$ is discontinuous. *Hint:* Consider $f(t) = \operatorname{sgn} t$, $[a, b] = [-1, 1]$.

5. Prove that, if $f(x)$ is infinitely differentiable over $[a, x]$, then

$$f(x) = f(a) + (x - a)f'(a) + \cdots + \frac{(x-a)^N}{N!} f^{(N)}(a) + R_N(x),$$

where

$$R_N(x) = \frac{1}{N!}\int_a^x (x - t)^N f^{(N+1)}(t)\,\mathrm{d}t.$$

(Taylor's theorem with integral form of remainder, cf. 5.41) *Hint:* Integrate $R_N(x)$ by parts N times.

6. Prove that if $f(x)$, $g(x)$ are continuous and $m \leqslant f(x) \leqslant M$, $g(x) \geqslant 0$ for all $x \in [a, b]$, then

$$m\int_a^b g(x)\,\mathrm{d}x \leqslant \int_a^b f(x)g(x)\,\mathrm{d}x \leqslant M\int_a^b g(x)\,\mathrm{d}x.$$

Deduce that

$$\int_a^b f(x)g(x)\,\mathrm{d}x = f(c)\int_a^b g(x)\,\mathrm{d}x$$

for some $c \in [a, b]$. (First mean value theorem for integrals.) *Hint:* Use 4.24.

7. Prove that if $f(x)$, $g(x)$ are continuous and

$$m \leqslant \int_a^x f(t)\,\mathrm{d}t \leqslant M,$$

$g(x) \geqslant 0$, $g'(x)$ exists and is continuous and $\leqslant 0$ for all $x \in [a, b]$, then

$$mg(a) \leqslant \int_a^b f(x)g(x)\,\mathrm{d}x \leqslant Mg(a)$$

(second mean value theorem for integrals).

Prove that, for any $0 < a < b$,

$$\left| \int_a^b \frac{\sin x}{x}\,\mathrm{d}x \right| \leqslant \frac{2}{a}.$$

8. Prove that, for any continuous $f(x)$, $g(x)$,

(i) $$\int_a^b |f(x) + g(x)|\,\mathrm{d}x \leqslant \int_a^b |f(x)|\,\mathrm{d}x + \int_a^b |g(x)|\,\mathrm{d}x,$$

(ii) $$\left(\int_a^b f(x)g(x)\,\mathrm{d}x\right)^2 \leqslant \int_a^b (f(x))^2\,\mathrm{d}x \int_a^b (g(x))^2\,\mathrm{d}x.$$

Hints: For (i) use 6.23 and 6.24. For (ii) consider $\int_a^b (f(x) + \lambda g(x))^2\,\mathrm{d}x$, where λ is a parameter.

9. Prove that, if $f(x)$ is continuous and $f(x) \geqslant 0$ for all $x \in [a, b]$, and if $f(c) > 0$ for some $c \in [a, b]$, then

$$\int_a^b f(x)\,\mathrm{d}x > 0.$$

Show by counter-example that we cannot drop the assumption that $f(x)$ is continuous.

10. By considering Riemann sums for suitable integrals, find the limits of the sequences whose nth terms are the following.

(i) $$\frac{1}{n+1} + \frac{1}{n+2} + \cdots + \frac{1}{2n}$$

(ii) $$\frac{1}{n+1} + \frac{1}{n+2} + \cdots + \frac{1}{3n}$$

(iii) $$\sqrt[n]{\left[\left(1+\frac{1}{n}\right)\left(1+\frac{2}{n}\right)\cdots\left(1+\frac{n}{n}\right)\right]}$$

7
Singular integrals

Up to now we have considered only integrals of bounded functions over bounded intervals. We now wish to allow the possibility of integrating unbounded functions over unbounded intervals. Many of the more interesting integrals of analysis come into this category: for example, the Laplace transform $F(x)$ of $f(t)$ is defined as

$$F(x)=\int_0^\infty e^{-xt}f(t)\,dt,$$

and the continuous factorial function $x!$ is defined as

$$x!=\int_0^\infty t^x e^{-t}\,dt.$$

We shall call integrals such as these *singular* integrals and classify them into three kinds. Singular integrals of the *first* kind are those which involve integration of a bounded function over an unbounded interval, e.g. $\int_1^\infty dx/x$. Singular integrals of the *second* kind will involve integration of an unbounded function over a bounded interval, e.g. $\int_0^1 dx/x$. Integrals of the *third* kind combine the characteristics of the other two kinds, e.g. $\int_0^\infty dx/x$.

We shall refer to points near which an integrand is unbounded as *singularities,* e.g. $\int_0^1 dx/x$ has a singularity at $x=0$. Integrals such as $\int_1^2 dx/x$, where there are no singularities and the range of integration is bounded will be called *ordinary* integrals.

Singular integrals will be defined by regarding them as limiting cases of ordinary integrals. We shall have to allow for the possibility of non-existence of limits by introducing the notions of convergence and divergence.

7.1 Definition

Suppose that $f(x)$ is continuous for all $x \geqslant a$, where a is fixed. We say $\int_a^\infty f(x)\,dx$ *converges,* or is *convergent,* if

$$\lim_{X\to\infty}\int_a^X f(x)\,dx$$

exists and is finite. When $\int_a^\infty f(x)\,\mathrm{d}x$ converges we define its value, also denoted by $\int_a^\infty f(x)\,\mathrm{d}x$, to be

$$\int_a^\infty f(x)\,\mathrm{d}x = \lim_{X\to\infty} \int_a^X f(x)\,\mathrm{d}x.$$

If $\int_a^\infty f(x)\,\mathrm{d}x$ fails to converge, we shall say it *diverges*, or is *divergent*. □

7.2 Example

$$\int_1^\infty \mathrm{d}x/x^\alpha \qquad (\alpha \text{ fixed}).$$

We have

$$\int_1^X \frac{\mathrm{d}x}{x^\alpha} = \left.\frac{x^{-\alpha+1}}{-\alpha+1}\right|_1^X$$

$$= \frac{X^{-\alpha+1}-1}{-\alpha+1}$$

$$\to \begin{cases} \dfrac{1}{\alpha-1} & \text{as } X\to\infty \text{ if } \alpha>1, \\ \infty & \text{as } X\to\infty \text{ if } \alpha<1. \end{cases}$$

If $\alpha = 1$,

$$\int_1^X \frac{\mathrm{d}x}{x} = \log x \Big|_1^X$$

$$= \log X$$

$$\to \infty \qquad \text{as } X\to\infty.$$

Hence, $\int_1^\infty \mathrm{d}x/x^\alpha$ converges if $\alpha > 1$, and diverges if $\alpha \leqslant 1$. For $\alpha > 1$ we have

$$\int_1^\infty \frac{\mathrm{d}x}{x^\alpha} = \frac{1}{\alpha - 1}.$$ □

7.3 Exercises

1. Which of the following integrals converge? Find the values of those which converge.

(i) $\int_1^\infty \mathrm{e}^{-\alpha x}\,\mathrm{d}x$ (ii) $\int_0^\infty \frac{\mathrm{d}x}{1+x}$ (iii) $\int_0^\infty \frac{\mathrm{d}x}{1+x^2}$

2. Show that if $f(x)$, $g(x)$ are continuous for $x \geqslant a$, and $\int_a^\infty f(x)\,\mathrm{d}x$, $\int_a^\infty g(x)\,\mathrm{d}x$ both converge, then also $\int_a^\infty (\alpha f(x) + \beta g(x))\,\mathrm{d}x$ converges for any constants α, β and its value is given by

$$\int_a^\infty (\alpha f(x) + \beta g(x))\,\mathrm{d}x = \alpha \int_a^\infty f(x)\,\mathrm{d}x + \beta \int_a^\infty g(x)\,\mathrm{d}x.$$

3. Show that, if $f(x)$ is continuous for $x \geqslant a$, and if $b > a$, then $\int_a^\infty f(x)\,\mathrm{d}x$ converges if and only if $\int_b^\infty f(x)\,\mathrm{d}x$ converges. □

7.4 Definition

Suppose that $f(x)$ is continuous for $a < x \leqslant b$, where $a < b$ are fixed. We say $\int_a^b f(x)\,\mathrm{d}x$ *converges*, or is *convergent*, if

$$\lim_{\varepsilon \to 0_+} \int_{a+\varepsilon}^b f(x)\,\mathrm{d}x$$

exists finite. When $\int_a^b f(x)\,\mathrm{d}x$ converges, we define its *value*, also denoted by $\int_a^b f(x)\,\mathrm{d}x$, to be

$$\int_a^b f(x)\,\mathrm{d}x = \lim_{\varepsilon \to 0_+} \int_{a+\varepsilon}^b f(x)\,\mathrm{d}x.$$

If $\int_a^b f(x)\,\mathrm{d}x$ fails to converge, we shall say it *diverges*, or is *divergent*. □

7.5 Example

$\int_0^1 \mathrm{d}x/x^\alpha$ (α fixed).

We have

$$\int_\varepsilon^1 \frac{\mathrm{d}x}{x^\alpha} = \left.\frac{x^{-\alpha+1}}{-\alpha+1}\right|_\varepsilon^1$$
$$= \frac{1-\varepsilon^{-\alpha+1}}{-\alpha+1}$$
$$\to \begin{cases} \dfrac{1}{1-\alpha} & \text{as } \varepsilon \to 0_+ \text{ if } \alpha < 1, \\ \infty & \text{as } \varepsilon \to 0_+ \text{ if } \alpha > 1. \end{cases}$$

If $\alpha = 1$,

$$\int_\varepsilon^1 \frac{\mathrm{d}x}{x} = \log x \Big|_\varepsilon^1$$
$$= -\log \varepsilon$$
$$\to \infty \qquad \text{as } \varepsilon \to 0_+.$$

Hence, $\int_0^1 \mathrm{d}x/x^\alpha$ converges if $\alpha<1$ and diverges if $\alpha\geqslant 1$. For $\alpha<1$ its value is

$$\int_0^1 \frac{\mathrm{d}x}{x^\alpha}=\frac{1}{1-\alpha}.$$

Observe that, for $\alpha\leqslant 0$, $\int_0^1 \mathrm{d}x/x^\alpha$ is an ordinary integral. □

7.6 Exercises

1. Say which integrals converge and find their values.

(i) $\displaystyle\int_0^1 \log x \,\mathrm{d}x$ (Integrate by parts)

(ii) $\displaystyle\int_0^{\frac{1}{2}\pi} \tan x \,\mathrm{d}x$

(iii) $\displaystyle\int_{-1}^1 \frac{\mathrm{d}x}{\sqrt{(1-x^2)}}$ (Substitute $x=\sin t$).

2. Show that if $f(x)$ is continuous over $a\leqslant x\leqslant b$, then

$$\lim_{\varepsilon\to 0+}\int_{a+\varepsilon}^b f(x)\,\mathrm{d}x=\int_a^b f(x)\,\mathrm{d}x$$

where $\int_a^b f(x)\,\mathrm{d}x$ is the ordinary integral. □

For integrals of the third kind we split the range of integration up to produce a sum of integrals of the first and second kind. We say they converge when *all* the component integrals converge, and the value is the sum of the values of the components.

7.7 Example

$\int_0^\infty \mathrm{d}x/x^\alpha$.

Write

$$\int_0^\infty \frac{\mathrm{d}x}{x^\alpha}=\int_0^1 \frac{\mathrm{d}x}{x^\alpha}+\int_1^\infty \frac{\mathrm{d}x}{x^\alpha}.$$

The first integral diverges for all $\alpha\geqslant 1$ (7.5), and the second diverges for all $\alpha\leqslant 1$ (7.2), so $\int_0^\infty \mathrm{d}x/x$ diverges for all α. □

We will give examples of convergent integrals of the third kind when we have developed a few tests for convergence.

The theory of convergence of singular integrals has many parallels with the theory of infinite series. We shall carry over much of the terminology and develop the theory in analogous fashion. For example, we shall begin by considering integrals with positive integrand. These are the analogues of series of positive terms. Later we define a notion of absolute convergence for integrals with general integrand. Also, we shall show there is a strong connection between integrals $\int_1^\infty f(x)\,dx$ of the first kind, and the corresponding series $\sum_1^\infty f(n)$.

7.8 Theorem

If $f(x) \geqslant 0$ and is continuous for all $x \geqslant a$, then $\int_a^\infty f(x)\,dx$ converges if and only if $\int_a^X f(x)\,dx$ is bounded for $X \geqslant a$.

Proof If we write $F(X) = \int_a^X f(x)\,dx$, then $F(X)$ is an increasing function of X, so converges to a finite limit or diverges to infinity according as it is bounded or not. □

The corresponding theorem for integrals of the second kind is that, if $f(x) \geqslant 0$ is continuous for $a < x \leqslant b$, then $\int_a^b f(x)\,dx$ converges if and only if $\int_{a+\varepsilon}^b f(x)\,dx$ is bounded for $\varepsilon > 0$ ($\varepsilon \leqslant b - a$).

7.9 Comparison test

If $0 \leqslant f(x) \leqslant g(x)$ are continuous for $x \geqslant a$, then convergence of $\int_a^\infty g(x)\,dx$ implies convergence of $\int_a^\infty f(x)\,dx$.

Proof If $\int_a^\infty g(x)\,dx$ converges, then, by 7.8, $\int_a^X g(x)\,dx$ is bounded for $X \geqslant a$, but therefore $\int_a^X f(x)\,dx$ is bounded for $X \geqslant a$, since

$$\int_a^X f(x)\,dx \leqslant \int_a^X g(x)\,dx,$$

and hence, by 7.8 again, $\int_a^\infty f(x)\,dx$ converges. □

7.10 Corollary

Under the same hypotheses, the divergence of $\int_a^\infty f(x)\,dx$ implies the divergence of $\int_a^\infty g(x)\,dx$. □

7.11 Examples

1. $\int_1^\infty dx/(1+x)$
   For all $x \geqslant 1$ we have

$$\frac{1}{1+x} \geqslant \frac{1}{2x}$$

and $\int_1^\infty dx/2x$ is divergent (see 7.2, 7.3). Hence $\int_1^\infty dx/(1+x)$ is divergent by the comparison test (7.10).

2. $\int_1^\infty dx/(1+x^2)$.
   For all $x \geqslant 1$ we have

$$\frac{1}{1+x^2} \leqslant \frac{1}{x^2}$$

and $\int_1^\infty dx/x^2$ is convergent (see 7.2). Hence $\int_1^\infty dx/(1+x^2)$ is convergent by the comparison test (7.9).

Both the above integrals can of course be treated by going back to the definition (see 7.3). □

7.12 Exercises

1. Discuss the convergence of the following integrals.

(i) $\displaystyle\int_1^\infty \frac{dx}{1+x^3}$

(ii) $\displaystyle\int_1^\infty e^{-x^2}\, dx \qquad (e^{-x^2} \leqslant e^{-x})$.

2. Show that, if $0 \leqslant f(x) \leqslant g(x)$ are continuous for $x \geqslant a$, then

$$\int_a^\infty f(x)\, dx \leqslant \int_a^\infty g(x)\, dx,$$

with the obvious interpretation if either integral diverges.

3. Formulate a comparison test for integrals of the second kind, and use it to discuss the convergence of the following integrals.

(i) $\displaystyle\int_0^{\frac{1}{2}\pi} \frac{dx}{\sin x} \qquad (\sin x \leqslant x)$

(ii) $\displaystyle\int_0^1 \frac{dx}{\log(1+x)} \qquad (\log(1+x) \leqslant x)$ □

7.13 Definition

Suppose $f(x)$ is continuous for $x \geqslant a$. We say the integral $\int_a^\infty f(x)\,\mathrm{d}x$ is *absolutely convergent*, or *converges absolutely*, if $\int_a^\infty |f(x)|\,\mathrm{d}x$ converges. □

7.14 Theorem

Absolute convergence implies convergence.

Proof Let

$$f^+(x) = \max\{f(x), 0\} = \tfrac{1}{2}(|f(x)| + f(x)),$$
$$f^-(x) = \max\{-f(x), 0\} = \tfrac{1}{2}(|f(x)| - f(x)).$$

Then $f^+(x)$, $f^-(x)$ are continuous and

$$f^+(x) \geqslant 0, \quad f^-(x) \geqslant 0,$$
$$f^+(x) - f^-(x) = f(x),$$
$$f^+(x) + f^-(x) = |f(x)|.$$

Therefore $f^+(x) \leqslant |f(x)|$, $f^-(x) \leqslant |f(x)|$ and so, by the comparison test, $\int_a^\infty f^+(x)\,\mathrm{d}x$, $\int_a^\infty f^-(x)\,\mathrm{d}x$ both converge. Hence $\int_a^\infty f(x)\,\mathrm{d}x$ converges by 7.3. □

7.15 Examples

$$\int_1^\infty \frac{\sin x}{x^2}\,\mathrm{d}x, \qquad \int_1^\infty \frac{\cos x}{x^2}\,\mathrm{d}x.$$

We have

$$\left|\frac{\sin x}{x^2}\right| \leqslant \frac{1}{x^2}$$

for all $x \geqslant 1$, so $\int_1^\infty (\sin x)/x^2\,\mathrm{d}x$ is absolutely convergent by the comparison test, and therefore convergent. Similarly for $\int_1^\infty (\cos x)/x^2\,\mathrm{d}x$. □

7.16 Exercises

1. Show that the following integrals are absolutely convergent and find their values.

$$\text{(i)} \int_0^\infty e^{-xt} \sin t \, dt \qquad \text{(ii)} \int_0^\infty e^{-xt} \cos t \, dt \qquad (x > 0)$$

2. Show that, if $\int_a^\infty f(x)\,dx$ is absolutely convergent, then

$$\left| \int_a^\infty f(x)\,dx \right| \leqslant \int_a^\infty |f(x)|\,dx.$$

3. Formulate a definition of absolute convergence for integrals of the second kind, and prove absolute convergence implies convergence.

□

7.17 Definition

$\int_a^\infty f(x)\,dx$ is *conditionally convergent* if convergent but not absolutely.

□

7.18 Example

$$\int_1^\infty \frac{\sin x}{x}\,dx \qquad \text{is conditionally convergent.}$$

Proof Integrating by parts we have

$$\int_1^X \frac{\sin x}{x}\,dx = -\frac{\cos x}{x}\bigg|_1^X - \int_1^X \frac{\cos x}{x^2}\,dx$$

$$= \cos 1 - \frac{\cos X}{X} - \int_1^X \frac{\cos x}{x^2}\,dx$$

$$\to \cos 1 - \int_1^\infty \frac{\cos x}{x^2}\,dx$$

as $X \to \infty$. (See 7.15.) Hence $\int_1^\infty (\sin x)/x\,dx$ is convergent.

However,

$$\left|\frac{\sin x}{x}\right| \geqslant \frac{\sin^2 x}{x} = \frac{1 - \cos 2x}{2x},$$

and

$$\int_1^X \frac{1 - \cos 2x}{2x}\,dx = \frac{1}{2}\int_1^X \frac{dx}{x} - \frac{1}{2}\int_1^X \frac{\cos 2x}{x}\,dx$$

$$\to \infty$$

as $X \to \infty$, since $\int_1^\infty (1/x)\,dx$ is divergent, and $\int_1^\infty (\cos 2x)/x\,dx$ is convergent (by the same argument as that used above to prove $\int_1^\infty (\sin x)/x \, dx$ is convergent). Hence, by the comparison test, $\int_1^\infty (\sin x)/x \, dx$ is not absolutely convergent.

7.19 The gamma function

We shall now illustrate the techniques of testing integrals of the third kind for convergence by giving a treatment of the integral $\int_0^\infty t^x e^{-t} \, dt$ for $x > -1$.

If $x \geqslant 0$, we have an integral of the first kind.

If $x < 0$, we have an integral of the third kind with a singularity at $t = 0$.

Consider $\int_1^\infty t^x e^{-t} \, dt$.

We have

$$t^x e^{-t} = (t^x e^{-\frac{1}{2}t})e^{-\frac{1}{2}t}$$

and $t^x e^{-\frac{1}{2}t} \to 0$ as $t \to \infty$; therefore there exists M such that

$$t^x e^{-\frac{1}{2}t} \leqslant M$$

for all $t \geqslant 1$. It follows that

$$t^x e^{-t} \leqslant M e^{-\frac{1}{2}t}$$

for all $t \geqslant 1$. Hence $\int_1^\infty t^x e^{-t} \, dt$ converges for all x by comparison with $\int_1^\infty e^{-\frac{1}{2}t} \, dt$ (see 7.3).

Now consider $\int_0^1 t^x e^{-t} \, dt$.

We have

$$t^x e^{-t} \leqslant t^x$$

for all $0 \leqslant t \leqslant 1$, and $\int_0^1 t^x \, dt$ converges for $x > -1$ (see 7.5). Hence $\int_0^1 t^x e^{-t} \, dt$ converges for all $x > -1$ by the comparison test.

Observe that, for any integer $n \geqslant 0$,

$$\begin{aligned} I_n &= \int_0^\infty t^n e^{-t} \, dt \\ &= -t^n e^{-t} \big|_0^\infty + n \int_0^\infty t^{n-1} e^{-t} \, dt \\ &= n I_{n-1} \\ &\;\;\cdots \\ &= n! \, I_0 \\ &= n! \end{aligned}$$

So it is natural to define

$$x! = \int_0^\infty t^x e^{-t} \, dt$$

for all real $x > -1$. Alternatively one defines the *gamma function*

$$\Gamma(x) = \int_0^{\infty} t^{x-1} e^{-t} \, dt$$

for $x > 0$, so that $\Gamma(n) = (n-1)!$ for all integers $n \geqslant 1$.

7.20 Exercise

Show $(-\frac{1}{2})! = 2\int_0^{\infty} e^{-t^2} \, dt (= \sqrt{\pi})$. □

7.21 Integral test (for series)

If $f(x) \geqslant 0$ is continuous and decreasing for $x \geqslant 1$, then convergence of the integral $\int_1^{\infty} f(x) \, dx$ is necessary and sufficient for convergence of the series $\sum_1^{\infty} f(n)$.

Proof We have

$$f(n+1) \leqslant f(x) \leqslant f(n)$$

for all $n \leqslant x \leqslant n+1$. Therefore

$$\int_n^{n+1} f(n+1) \, dx \leqslant \int_n^{n+1} f(x) \, dx \leqslant \int_n^{n+1} f(n) \, dx,$$

i.e.

$$f(n+1) \leqslant \int_n^{n+1} f(x) \, dx \leqslant f(n).$$

Therefore, if we sum over $n = 1, 2, \ldots, N$, we obtain

$$\sum_2^{N+1} f(n) \leqslant \int_1^{N+1} f(x) \, dx \leqslant \sum_1^{N} f(n).$$

It follows that $\sum_1^N f(n)$ is bounded over N if and only if $\int_1^X f(x) \, dx$ is bounded over X. Hence the series $\sum_1^{\infty} f(n)$ converges if and only if the integral $\int_1^{\infty} f(x) \, dx$ does. □

7.22 Example

$\sum_1^{\infty} (1/n^{\alpha})$.

The function $f(x) = 1/x^{\alpha}$ satisfies the conditions of 7.21 if $\alpha \geqslant 0$. Therefore, by 7.2, we find $\sum_1^{\infty} (1/n^{\alpha})$ is convergent if $\alpha > 1$, and divergent if $0 \leqslant \alpha \leqslant 1$. In fact, $\sum_1^{\infty} (1/n^{\alpha})$ diverges for all $\alpha \leqslant 1$, since, for $\alpha < 0$, the nth term $1/n^{\alpha} \not\to 0$. □

7.23 Euler's limit

The sequence (γ_n), where

$$\gamma_n = 1 + \tfrac{1}{2} + \cdots + \frac{1}{n} - \log n$$

is decreasing and satisfies $0 \leqslant \gamma_n \leqslant 1$ for all n, hence is convergent to a finite limit γ, also satisfying $0 \leqslant \gamma \leqslant 1$, as $n \to \infty$. γ is known as Euler's constant and its value is 0.577 to 3 decimal places.

To justify these assertions, we argue as in the proof of 7.21 for the particular case $f(x) = 1/x$. We arrive at the inequalities

$$\sum_{2}^{N+1} \frac{1}{n} \leqslant \int_{1}^{N+1} \frac{\mathrm{d}x}{x} \leqslant \sum_{1}^{N} \frac{1}{n},$$

which give

$$(\log N \leqslant)\, \log (N+1) \leqslant \sum_{1}^{N} \frac{1}{n} \leqslant 1 + \log N$$

and therefore $0 \leqslant \gamma_N \leqslant 1$. To show γ_n decreases, we observe that

$$\begin{aligned}\gamma_n - \gamma_{n+1} &= -\frac{1}{n+1} - \log n + \log (n+1) \\ &= \int_{n}^{n+1} \left(\frac{1}{x} - \frac{1}{n+1}\right) \mathrm{d}x \\ &\geqslant 0,\end{aligned}$$

since the integrand is positive. □

7.24 Estimates of $\sum_1^N (1/n)$

The above analysis can be used to give us some idea of how large the partial sums of the harmonic series $\sum_1^\infty (1/n)$ are. We know this series diverges to infinity. We shall see, however, that it does so remarkably slowly.

In fact,

$$\begin{aligned}\sum_{1}^{N} \frac{1}{n} &= \gamma_N + \log N \\ &\leqslant 1 + \log N.\end{aligned}$$

Now $\log 10 = 2.3026$ to 4 decimal places, so e.g. $\sum_1^{100} (1/n) < 6$,

$\sum_1^{10^6}(1/n)<15$. On the other hand, we also have

$$\sum_1^N \frac{1}{n} \geqslant \gamma + \log N,$$

which shows that, if e.g. we want $\sum_1^N (1/n) > 100$, we require N to be in the region of $e^{100-\gamma} = 1.51 \times 10^{43}$ to 3 significant figures.

7.25 Rearrangements of $\sum_1^\infty (-1)^{n-1}/n$

We can use Euler's limit to find the sum of the alternating harmonic series $\sum_1^\infty (-1)^{n-1}/n$, and various rearrangements of it, as follows.

$$\begin{aligned}\sum_1^{2N} (-1)^{n-1}/n &= 1 - \tfrac{1}{2} + \tfrac{1}{3} - \tfrac{1}{4} + \cdots - \frac{1}{2N} \\ &= \sum_1^{2N} (1/n) - \sum_1^{N} (1/n) \\ &= (\log 2N + \gamma_{2N}) - (\log N + \gamma_N) \\ &= \log 2 + \gamma_{2N} - \gamma_N \\ &\to \log 2\end{aligned}$$

as $N \to \infty$. Hence $\sum_1^\infty (-1)^{n-1}/n = \log 2$.

Consider e.g. the rearrangement

$$1 + \tfrac{1}{3} - \tfrac{1}{2} + \tfrac{1}{5} + \tfrac{1}{7} - \tfrac{1}{4} + \cdots$$

The $3N$th partial sum is

$$\begin{aligned}&1 + \tfrac{1}{3} - \tfrac{1}{2} + \tfrac{1}{5} + \tfrac{1}{7} - \tfrac{1}{4} + \cdots + \frac{1}{4N-3} + \frac{1}{4N-1} - \frac{1}{2N} \\ &- \sum_1^{4N} (1/n) - \tfrac{1}{2}\sum_1^{2N} (1/n) - \tfrac{1}{2}\sum_1^{N} (1/n) \\ &= (\log 4N + \gamma_{4N}) - \tfrac{1}{2}(\log 2N + \gamma_{2N}) - \tfrac{1}{2}(\log N + \gamma_N) \\ &= \tfrac{3}{2}\log 2 + \gamma_{4N} - \tfrac{1}{2}\gamma_{2N} - \tfrac{1}{2}\gamma_N \\ &\to \tfrac{3}{2}\log 2\end{aligned}$$

as $N \to \infty$. Hence

$$1 + \tfrac{1}{3} - \tfrac{1}{2} + \tfrac{1}{5} + \tfrac{1}{7} - \tfrac{1}{4} + \cdots = \tfrac{3}{2}\log 2. \qquad \square$$

7.26 Miscellaneous exercises

1. Discuss the convergence of the following integrals.

(i) $\displaystyle\int_1^\infty \frac{\sin x}{x^\alpha}\,dx \quad (\alpha > 0)$ (ii) $\displaystyle\int_1^\infty \sin(x^2)\,dx$

(iii) $\int_1^\infty x \sin(x^3)\, dx$ (iv) $\int_0^\infty \frac{\sin x}{x^{3/2}}\, dx$

(v) $\int_0^1 \frac{1}{x} \sin\left(\frac{1}{x}\right) dx$ (vi) $\int_0^{\frac{1}{2}\pi} \sec x\, dx$

(vii) $\int_0^\pi \log \sin x\, dx$ (viii) $\int_0^1 \frac{\log x}{x}\, dx$

2. Does $\int_1^\infty f(x)\, dx$ convergent imply $f(x) \to 0$ as $x \to \infty$? In general? If $f(x)$ is decreasing? If $f(x) \geqslant 0$?

3. Show that

$$\begin{aligned}\int_0^{\frac{1}{2}\pi} \log \sin x\, dx &= \int_0^{\frac{1}{2}\pi} \log \cos x\, dx \qquad (u = \tfrac{1}{2}\pi - x)\\ &= \tfrac{1}{2}\int_0^{\frac{1}{2}\pi} \log(\tfrac{1}{2}\sin 2x)\, dx\\ &= -\tfrac{1}{2}\pi \log 2.\end{aligned}$$

4. Show that

$$\int_0^{2\pi} \log |1 - e^{i\theta}|\, d\theta = 0.$$

5. Show that

$$\begin{aligned}\int_0^\infty \frac{dx}{1+x^3} &= \int_0^\infty \frac{x\, dx}{1+x^3} \qquad (x = 1/u)\\ &= \tfrac{1}{2}\int_0^\infty \frac{dx}{1 - x + x^2}\\ &= 2\pi/3\sqrt{3}.\end{aligned}$$

6. Show

(i) $$\int_0^\infty \frac{dx}{1+x^4} = \int_0^\infty \frac{x^2\, dx}{1+x^4} \qquad (x = 1/u)$$

(ii) $$\int_0^\infty \frac{x\, dx}{1+x^4} = \tfrac{1}{2}\int_0^\infty \frac{dx}{1+x^2} = \frac{\pi}{4} \qquad (u = x^2)$$

(iii) $$\begin{aligned}\int_0^\infty \frac{dx}{1+x^4} &= \tfrac{1}{2}\int_0^\infty \frac{1+x^2}{1+x^4}\, dx\\ &= \tfrac{1}{2}\int_0^\infty \frac{dx}{1+\sqrt{2}x + x^4} + \frac{1}{\sqrt{2}}\int_0^\infty \frac{x\, dx}{1+x^4}\\ &= \pi/2\sqrt{2}.\end{aligned}$$

7. Consider the series $\sum_2^\infty 1/n^\alpha(\log n)^\beta$.

(i) Show that, if $\alpha > 1$, it converges for all β. *Hint* Let $\alpha = 1 + \varepsilon$ ($\varepsilon > 0$) and write

$$\frac{1}{n^\alpha(\log n)^\beta} = \frac{1}{n^{1+\frac{1}{2}\varepsilon}} \frac{1}{n^{\frac{1}{2}\varepsilon}(\log n)^\beta}$$

and observe that $1/n^{\frac{1}{2}\varepsilon}(\log n)^\beta \to 0$ and $\sum_1^\infty 1/n^{1+\frac{1}{2}\varepsilon}$ converges.

(ii) Show that, if $\alpha < 1$, it diverges for all β. *Hint* Let $\alpha = 1 - \varepsilon$ ($\varepsilon > 0$) and argue similarly.

(iii) Show that, if $\alpha = 1$, it converges for all $\beta > 1$ and diverges for all $\beta \leqslant 1$. *Hint* Use the integral test (7.21).

8. Show that $\sum_{N+1}^\infty (1/n^2) \leqslant 1/N$. *Hint* Observe that

$$\int_{n-1}^{n} \frac{dx}{n^2} \leqslant \int_{n-1}^{n} \frac{dx}{x^2}.$$

9. Let $\gamma_n = 1 + \frac{1}{2} + \cdots + \dfrac{1}{n} - \log n$. Show that

$$\gamma_n - \gamma_{n+1} = \frac{1}{n+1} \int_0^1 \frac{1-t}{n+t}\, dt.$$

Deduce that

(i) $\gamma_n - \gamma_{n+1} \leqslant \dfrac{1}{2n^2}$,

(ii) $\gamma_n - \gamma \leqslant \dfrac{1}{2(n-1)}$,

where $\gamma = \lim_{n\to\infty} \gamma_n$.

10. Show that $\sum_1^{1000} (1/n) = 6.908$ to 3 decimal places. (Assume $\log 10 = 2.3026$ to 4 decimal places.)

11. Use Euler's limit to show that

(i) $1 - \frac{1}{2} - \frac{1}{4} + \frac{1}{3} - \frac{1}{6} - \frac{1}{8} + \cdots = \frac{1}{2} \log 2$.

(ii) $1 - \frac{1}{2} + \frac{1}{3} + \frac{1}{5} - \frac{1}{4} + \frac{1}{7} + \frac{1}{9} + \frac{1}{11} - \frac{1}{6} + \cdots$ diverges to infinity.

12. By considering Riemann sums for suitable singular integrals, find the limits of the following sequences.

(i) $\dfrac{(n!)^{1/n}}{n}$

(ii) $\dfrac{1}{\sqrt{n}} + \dfrac{1}{\sqrt{[2(n-1)]}} + \dfrac{1}{\sqrt{[3(n-2)]}} + \cdots + \dfrac{1}{\sqrt{n}}$

(cf. 3.38, question 12).

Appendix

1. Euler's proof that $\sum_1^\infty 1/n^2 = \pi^2/6$

The proof depends on the fact that, if the roots of the polynomial equation

$$a_0 + a_1 x + \cdots + a_n x^n = 0$$

are $\alpha_1, \alpha_2, \ldots, \alpha_n$, then

$$\frac{1}{\alpha_1} + \frac{1}{\alpha_2} + \cdots + \frac{1}{\alpha_n} = -\frac{a_1}{a_0}.$$

Now the roots of the equation

$$\frac{\sin x}{x} = 1 - \frac{x^2}{3!} + \frac{x^4}{5!} - \cdots = 0$$

are $\pm\pi, \pm 2\pi, \ldots, \pm n\pi, \ldots$, and so the roots of

$$1 - \frac{x}{3!} + \frac{x^2}{5!} - \cdots = 0$$

are $\pi^2, 4\pi^2, \ldots, n^2\pi^2, \ldots$. Hence

$$\frac{1}{\pi^2} + \frac{1}{4\pi^2} + \cdots + \frac{1}{n^2\pi^2} + \cdots = \frac{1}{3!},$$

which gives

$$\sum_1^\infty \frac{1}{n^2} = \frac{\pi^2}{6}.$$

□

2. Proof that $\sum_1^\infty \sin nx/n = \frac{1}{2}(\pi - x)$ for all $0 < x < 2\pi$

We proved in Chapter 5 that

$$\log(1+x) = x - \frac{x^2}{2} + \frac{x^3}{3} - \frac{x^4}{4} + \cdots$$

for all $-1<x\leqslant 1$ (see 5.22 and 5.42). If we put $x=-e^{i\theta}$, where $0<\theta<2\pi$, we get

$$\log(1-e^{i\theta}) = -e^{i\theta} - \frac{e^{2i\theta}}{2} - \frac{e^{3i\theta}}{3} - \frac{e^{4i\theta}}{4} - \cdots.$$

Taking imaginary parts gives

$$\begin{aligned}\sum_1^\infty \frac{\sin n\theta}{n} &= -\arg(1-e^{i\theta})\\ &= \frac{\pi-\theta}{2},\end{aligned}$$

since, e.g.

$$\begin{aligned}1-e^{i\theta} &= -e^{\frac{1}{2}i\theta}(e^{\frac{1}{2}i\theta} - e^{-\frac{1}{2}i\theta})\\ &= -e^{\frac{1}{2}i\theta}(2i\sin\tfrac{1}{2}\theta)\\ &= 2\sin\tfrac{1}{2}\theta e^{\frac{1}{2}i(\theta-\pi)}.\end{aligned}$$

□

Observe that taking real parts gives

$$\sum_1^\infty \frac{\cos n\theta}{n} = -\log(2\sin\tfrac{1}{2}\theta)$$

for all $0<\theta<2\pi$.

Solutions to exercises

Chapter 1

1.7 $x > 16.25$.

1.9 $x < -3$ or >2.

1.18 (i) $-2 < x < 0$. (ii) $x < 0$.

1.34 1. (i) $-2.5 < x < -2$. (ii) $x < -4$ or > -2. (iii) $-1.5 < x < -1$. (iv) $1.5 < x < 4$.

6. Observe that $M - \varepsilon$ cannot be an upper bound.

Chapter 2

2.2 (i) Take $N \geqslant 1/\sqrt{\varepsilon}$. (ii) Take $N \geqslant \log_2 1/\varepsilon$. (iii) Take $N \geqslant 1/\varepsilon^2$.

2.13 (i) 1. (ii) 0. (iii) -1.

2.21 (i) Take $N \geqslant \sqrt{C}$. (ii) Take $N \geqslant C^2$. (iii) Take $N \geqslant \log_2 C$.

2.32 (i) Not monotonic. (ii) Is monotonic (strictly increasing).

2.39 1. (i) Converges to 1. (ii) Diverges to infinity. (iii) Diverges to infinity. Use Bernoulli's inequality and the open sandwich principle. (iv) Converges to 1. Use Bernoulli and the sandwich principle. (v) Sequence is $\leqslant (\frac{5}{6})^n$ for all $n \geqslant 3$, therefore is null. (vi) Is null. Divide through by 4^n. (vii) Diverges to infinity. (viii) Oscillates. (ix) Oscillates.

3. Last part. Write $\sqrt{(n+1)} - \sqrt{n} = 1/[\sqrt{(n+1)} + \sqrt{n}]$.

5. The converse requires $a_n > 0$.

9. For each positive integer n there exists $a_n \in E$ such that $a_n > n$.

10. Choose a subsequence (a_{n_r}) which converges to a. (a_n) cannot converge to a so there must exist $\varepsilon > 0$ such that for any N we must have $|a_n - a| \geqslant \varepsilon$ for some $n > N$. This enables us to choose a subsequence (a_{m_r}) such that $|a_{m_r} - a| \geqslant \varepsilon$ for all r. Now choose a convergent subsequence of (a_{m_r}). It must converge to $b \neq a$.

11. $a = \sqrt{A}$.

12. $l = \sqrt{2}$.

Chapter 3

3.3 $\sum_1^\infty 1/n(n+1) = 1$.

3.23 (i) Convergent. (ii) Divergent.

3.29 (i) Conditionally convergent. (ii) Absolutely convergent.

3.32 (i) Converges absolutely for all x. (ii) Converges absolutely for $|x| < 1$, conditionally for $x = \pm 1$, diverges for $|x| > 1$.

3.34 (i) $\frac{1}{3}$. (ii) $\frac{1}{4}$.

3.38 1. They all converge except (ii) and (vi).

2. (i) Absolutely convergent. (ii) Conditionally convergent. (iii) Divergent. (iv) Conditionally convergent. (v) Conditionally convergent. (vi) Absolutely convergent.

3. (i) $\frac{3}{2}$. (ii) 1. (iii) 1. (iv) 4/27. (v) 1/e. (vi) 2.

5. Limit of the sequence is 0. Radius of convergence of the power series is 2.

6. The series is divergent. Observe, e.g., that

$$\frac{1}{\sqrt{(2n-1)}} - \frac{1}{2n} \geqslant \frac{1}{2\sqrt{(2n-1)}}$$

for all $n \geqslant 1$, and deduce that the even partial sums diverge to infinity.

7. $\sum_1^\infty 1/n^\alpha$ converges for $\alpha > 1$, diverges for $\alpha \leqslant 1$.

8. Nothing can be said if $l = 1$.

9. The power series diverges at $x = \frac{1}{4}$ by use of this comparison test with the series $\sum_1^\infty 1/n$, and converges at $x = -\frac{1}{4}$ by the alternating series test. (To show the sequence of absolute values is null, compare the ratio of consecutive terms with that for $n^{-1/3}$.)

16. If the nth partial sum of $\sum_1^\infty a_n$ is s_n and its sum to infinity is s, then the nth partial sum of $\sum_{N+1}^\infty a_n$ is $s_{N+n} - s_N$, which converges to $s - s_N$.

17. If $|x| < R$, choose r satisfying $|x| < r < R$, write $na_n x^{n-1} = n(x/r)^{n-1} a_n r^{n-1}$ and observe that $\sum_1^\infty n(x/r)^{n-1}$ is absolutely convergent by the ratio test, and $(a_n r^{n-1})_{n \geqslant 1}$ is a null sequence. If $|x| > R$, show $\sum_1^\infty na_n x^{n-1}$ is not absolutely convergent by comparison with $\sum_1^\infty a_n x^n$.

Chapter 4

4.17 (i) No limit. (ii) Limit exists = 0. (iii) No limit.

4.59 1. (i) Discontinuities at $x = n\pi$. (ii) Discontinuity at $x = 0$. (iii) Function is $|x|$ which is continuous everywhere. (iv) Discontinuities at $x = 1/n\pi$ and $x = 0$. (v) Continuous everywhere. (vi) Discontinuous everywhere. (vii) Continuous at $x = \frac{1}{2}$ only.

2. (i) 1. (ii) 1. (iii) $\frac{1}{2}$.

4. To prove the left-hand inequality, observe that the left-hand inequality of question 3 is true for all $x \geqslant 0$ (clearly from the definition of e^x), therefore the right-hand inequality of question 3 is true for all $x < 1$ (put $x = -u$ for $x < 0$); now put

$$x = \frac{u}{1+u} \qquad (u > -1)$$

and take logarithms. (Draw diagrams!)

6. The series converges conditionally. Use alternating series test. Show $(\log n)/n$ decreases for $n \geqslant 3$, e.g. by showing $(n+1)^n < n^{n+1}$ for $n \geqslant 3$ by 2.34.

7. Compare the Maclaurin series of the moduland with a geometric series.

Chapter 5

$$\text{5.12} \qquad f'(x) = 2x \sin\frac{1}{x} - \cos\frac{1}{x} \qquad (x \neq 0),$$
$$= 0 \qquad (x = 0).$$

Use 5.6 for $x \neq 0$ and 5.1 for $x = 0$. Theorem 5.3 says $f(x)$ is continuous, *not* $f'(x)$ continuous.

5.26 Maximum at $x = -2$, minimum at $x = 0$.

5.32 (i) 1. (ii) -1. (iii) $\frac{1}{2}$.

5.36 See remarks after 5.37.

5.39 (i) $e^x = e^2 e^{x-2}$. (ii) $\sin x = \cos(x - \frac{1}{2}\pi)$. (iii) $\Sigma_0^\infty (x+1)^n / 2^{n+1}$.

5.44 1. (i) Discontinuous. (ii) Continuous, not differentiable. (iii) Differentiable.

2. (i) Maximum at $x = 0$. (ii) Minimum at $x = -1/\sqrt{2}$, maximum at $x = 1/\sqrt{2}$. (iii) Maxima at $x = \pm 1$, minimum at $x = 0$.

3. (ii) Differentiate

$$f(x) = \sin x - \frac{x(\pi - x)}{\pi}$$

twice. (iii) Observe that

$$\frac{\sin x}{x} \geqslant 1 - \frac{x^2}{6}$$

for all x, and

$$1-\frac{x^2}{6}\geqslant\frac{\pi^2-x^2}{\pi^2+x^2}$$

if $x^2\leqslant 12-\pi^2$. If we put $x=\pi-u$ the inequality is equivalent to

$$\frac{\sin u}{u}\geqslant\frac{2\pi^2-3\pi u+u^2}{2\pi^2-2\pi u+u^2}$$

if $0\leqslant u\leqslant\pi$. By the same method this inequality is true if

$$6\pi-2\pi^2u+2\pi u^2-u^3\geqslant 0$$

which certainly holds for $0\leqslant u\leqslant 2$. If we put $x=\pi+u$ the inequality is equivalent to

$$\frac{\sin u}{u}\leqslant\frac{2\pi^2+3\pi u+u^2}{2\pi^2+2\pi u+u^2}$$

if $u\geqslant 0$, and this holds for all $u\geqslant 0$ since the right-hand side is $\geqslant 1$.

4. (i) 2. (ii) $\frac{1}{2}$. (iii) log 2. (iv) e^3.

5. Use L'Hôpital's rule (differentiating with respect to h). The second limit is $-f'''(x)$, on the assumption that $f(x)$ is thrice differentiable at x. The general case involves bionomial coefficients.

6. Consider

$$\frac{f'(x)-f'(c)}{x-c}$$

for x near c and use 5.24. $f''(c)<0$ gives a local maximum at $x=c$. No conclusion can be drawn in case $f''(c)=0$.

12. Equality is at $x=0$ only ($\alpha\neq 0, 1$).

Chapter 6

6.2

$$L_D=\frac{N-1}{2N},\ U_D=\frac{N+1}{2N}.$$

6.16 All three functions are integrable, (i) since monotonic, (ii) since continuous, (iii) since for any D including $1\pm\varepsilon$, $2\pm\varepsilon$ we have $U_D-L_D\leqslant 4\varepsilon$.

6.33 (i) -2. (ii) $e-2$. (iii) $\frac{1}{2}(1+e^{\pi})$. (iv) $2\log 2-1$.

6.36 (i) $\pi/6$. (ii) $2-2\log 2$. (iii) 1. (iv) = (ii).

6.37 1. (i) Given $\varepsilon>0$, choose D as follows. Take $x_1=\varepsilon$. Choose

$x_2, \ldots, x_N$ to give a dissection D' of $[\varepsilon, 1]$ such that $U_{D'} - L_{D'} < \varepsilon$. ($\sin 1/x$ is integrable over $[\varepsilon, 1]$ since continuous on this interval.) We obtain $U_D - L_D < 3\varepsilon$ which shows $\sin 1/x$ is integrable over $[0, 1]$ by 6.13. (ii) This function is continuous therefore integrable.

4. If $|f(t)| \leqslant M$ for all $t \in [a, b]$, then

$$|F(x+h) - F(x)| \leqslant \int_x^{x+h} |f(t)|\, \mathrm{d}t \leqslant hM$$

if $h > 0$.

7. Write $F(x) = \int_a^x f(t)\, \mathrm{d}t$ and observe that

$$\int_a^b f(x)g(x)\, \mathrm{d}x = F(b)g(b) - \int_a^b F(x)g'(x)\, \mathrm{d}x.$$

Now use question 6. For the last part put $f(x) = \sin x$, $g(x) = 1/x$.

9. By continuity there exists $\delta > 0$ such that $f(x) \geqslant \frac{1}{2}f(c)$ for all $|c - x| \leqslant \delta$. Therefore

$$\int_a^b f(x)\, \mathrm{d}x \geqslant \int_{c-\delta}^{c+\delta} \tfrac{1}{2}f(c)\, \mathrm{d}x = \delta f(c) > 0.$$

See 6.11 (ii) for the last part.

10. (i) $\int_1^2 \mathrm{d}x/x = \log 2$. (ii) $\int_1^3 \mathrm{d}x/x = \log 3$. (iii) The logarithm of the sequence converges to $\int_1^2 \log x\, \mathrm{d}x = 2\log 2 - 1$, therefore the sequence itself converges to $4/\mathrm{e}$.

Chapter 7

7.3 1. (i) Converges only if $\alpha > 0$ when its value is $\mathrm{e}^{-\alpha}/\alpha$. (ii) Diverges. (iii) Converges with value $\frac{1}{2}\pi$.

7.6 1. (i) Convergent with value -1. (ii) Divergent. (iii) Convergent with value π.

2. Use question 4 of 6.37.

7.12 1. (i) Convergent by comparison with $\int_1^\infty \mathrm{d}x/x^3$ (see 7.2).

7.16 1. (i) $1/(1+x^2)$. (ii) $x/(1+x^2)$.

2. Use 6.25.

7.20 Make the substitution $t = x^2$ in the integral $\int_0^\infty t^{-\frac{1}{2}}e^{-t}\, \mathrm{d}t$.

7.26 1. (i) Conditionally convergent for $0 < \alpha \leqslant 1$, absolutely convergent for $\alpha > 1$. Use the methods of 7.15, 7.18. (ii) Conditionally convergent. Substitute $x = \sqrt{t}$. (iii) Conditionally convergent. Substitute $x = \sqrt[3]{t}$. (iv) Convergent. Compare with $\int_0^1 x^{-\frac{1}{2}}\, \mathrm{d}x$ at $x = 0$. (v) Conditionally convergent. Put $x = 1/t$. (vi) Divergent. (vii) Convergent. Compare with $\int_0^1 \log x\, \mathrm{d}x$ at $x = 0$, $\int_0^\pi \log(\pi - x)\, \mathrm{d}x$ at $x = \pi$. (viii) Divergent. The indefinite integral of $(\log x)/x$ is $\frac{1}{2}(\log x)^2$.

2. No in general. Consider question 1, (ii) and (iii). Yes if $f(x)$ is decreasing. Observe that

$$0 \leqslant f(X) \leqslant \int_{X-1}^{X} f(x)\,\mathrm{d}x \to 0$$

as $X \to \infty$. No if $f(x) \geqslant 0$, consider e.g. $f(x) = 1 - 2^n\,|x - n|$ for x satisfying $|x - n| \leqslant 1/2^n$ and otherwise zero.

4. Use question 3.

9. (ii) Use question 8.

10. Use question 9 and the approximate value of γ given in 5.23.

11. Proceed as in 7.25.

12. (i) The logarithm of the sequence converges to $\int_0^1 \log x\,\mathrm{d}x = -1$. Therefore the sequence itself converges to 1/e. To justify this assertion observe that

$$\int_0^1 \log x\,\mathrm{d}x < \frac{1}{n}\log\left(\frac{1}{n}\frac{2}{n}\cdots\frac{n}{n}\right) < \int_{1/n}^1 \log x\,\mathrm{d}x$$

and use the sandwich principle. (ii) This sequence converges to $\int_0^1 \mathrm{d}x/\sqrt{[x(1-x)]} = \pi$. To justify observe

$$\int_{1/(n+1)}^{n/(n+1)} \frac{\mathrm{d}x}{\sqrt{[x(1-x)]}} < \frac{1}{\sqrt{n}} + \cdots + \frac{1}{\sqrt{n}} < \int_0^1 \frac{\mathrm{d}x}{\sqrt{[x(1-x)]}}.$$

(Draw diagrams.)

Index of symbols

Index